AESTHETICS AND THE ENVIRONMENT

Aesthetics and the Environment: The Appreciation of Nature, Art and Architecture presents fresh and fascinating insights into our interpretation of the environment. Although traditional aesthetics is often associated with the appreciation of art, Allen Carlson shows how much of our aesthetic experience does not encompass art but nature, in our responses to sunsets, mountains, or more mundane surroundings, such as gardens or the views from our windows. He demonstrates that, unlike works of art, natural and ordinary human environments are neither self-contained aesthetic objects nor specifically designed for convenient aesthetic consumption. On the contrary, our environments are ever present constantly engaging our senses, and Carlson offers a thought-provoking and lucid investigation of what this means for our appreciation of the world around us. He argues that knowledge of what it is we are appreciating is essential to having an appropriate aesthetic experience and that scientific understanding of nature can enhance our appreciation of it, rather than denigrate it.

Aesthetics and the Environment also shows how ethical and aesthetic values are closely connected, argues that aesthetic appreciation of natural and human environments has objective grounding, and explores the important links between ecology and the aesthetic experience of nature. This book will be essential reading for those involved in environmental studies and aesthetics and all who are interested in the controversial relationship between science and nature.

Allen Carlson is an authority in aesthetics and has pioneered the field of environmental aesthetics. He is Professor of Philosophy at the University of Alberta, Canada.

AESTHETICS AND THE ENVIRONMENT

The appreciation of nature, art and architecture

Allen Carlson

London and New York

First published 2000
by Routledge
11 New Fetter Lane, London EC4P 4EE

Simultaneously published in the USA and Canada
by Routledge
29 West 35th Street, New York, NY 10001

Routledge is an imprint of the Taylor & Francis Group

Typeset in Times
by HWA Text and Data Management, Tunbridge Wells
Printed and bound in Great Britain
by TJ International Ltd, Padstow, Cornwall

British Library Cataloguing in Publication Data
A catalogue record for this book is available
from the British Library

Library of Congress Cataloging in Publication Data
Carlson, Allen
Aesthetics and the environment: the appreciation of nature, art and architecture / Allen Carlson
p. cm.
"Simultaneously published in the USA and Canada"
Includes bibliographical references and index
1. Environment (Aesthetics)
BH301.E58C37 1999 99-28959
111'.85–dc21 CIP

ISBN 0-415-20683-9

FOR ARLENE, ELEANOR AND VIRGINIA

CONTENTS

ILLUSTRATIONS

PREFACE

The material in this volume spans a period of roughly twenty years and reflects my continuing interest in the aesthetics of the environment.

A number of the chapters of the volume have been previously published as essays in books or journals. Except for slight changes required to make corrections, eliminate redundancies, or indicate connections, these essays are reprinted much as they initially appeared. I have elected this alternative both because it best maintains the integrity and quality of the individual pieces and because I think that even in their original form, the essays come together to constitute a unified line of thought. The introductory chapters for the two parts of the volume, Chapters 1 and 8, present overviews and help to further unify the material. Moreover, the notes for these two chapters provide extensive cross-referencing among the remaining chapters and position them within the current literature in the field. In addition to Chapters 1 and 8, two other chapters, Chapters 2 and 14, have not previously appeared in print.

Allen Carlson
Edmonton, Canada, 1998

ACKNOWLEDGMENTS

I acknowledge and thank the editors and the publishers of the following books and journals for permission to use material that originally appeared in them: For portions of the general Introduction and of Chapter 1, by permission of, respectively, Basil Blackwell, Routledge, and Oxford University Press, D. Cooper (ed.) *A Companion to Aesthetics*, Oxford, Basil Blackwell, 1992, pp. 142–4; E. Craig (ed.) *The Routledge Encyclopedia of Philosophy*, London, Routledge, 1998, vol. 6, pp. 731–5; and M. Kelly (ed.) *The Encyclopedia of Aesthetics*, New York, Oxford University Press, 1998, vol. 3, pp. 346–9. For Chapters 3 and 9, by permission of the University of Illinois Press, *The Journal of Aesthetic Education*, 1979, vol. 13, pp. 99–114 and 1976, vol. 10, pp. 69-82. For Chapters 4, 5, and 12, by permission of the University of Wisconsin Press, *The Journal of Aesthetics and Art Criticism*, 1979, vol. 37, pp. 267–76; 1981, vol. 40, pp. 15–27, and 1985, vol. 43, pp. 301–12. For Chapter 6, by permission of Environmental Philosophy, Inc., *Environmental Ethics*, 1984, vol. 6, pp. 5–34. For Chapter 7, by permission of Cambridge University Press, S. Kemal and I. Gaskell (eds) *Landscape, Nature Beauty and the Arts*, Cambridge, Cambridge University Press, 1993, pp. 199–227. For Chapter 10, by permission of the University of Calgary Press, *The Canadian Journal of Philosophy*, 1986, vol. 16, pp. 635–50. For Chapter 11, by permission of Oxford University Press, *The British Journal of Aesthetics*, 1997, vol. 37, pp. 47–56. For Chapter 13, by permission of Editions Rodopi B.V., M. Mitias (ed.) *Philosophy and Architecture*, Amsterdam, Rodopi, 1994, pp. 141–64.

I also express my appreciation to the following by the courtesy of whom it was possible to include illustrations 2, 3, 4, 8, and 9: respectively, Corbis/ Ansel Adams Publishing Rights Trust, The John Weber Gallery, Alan Sonfist, The National Gallery of Canada, and American Telephone and Telegraph.

I thank the Social Sciences and Humanities Research Council of Canada and the University of Alberta for financial support for the writing of some of the material in this volume.

I also thank my editors at Routledge for their assistance and advice.

Above all, I thank and express my appreciation to my friends and acquaintances in the discipline, to my colleagues in the Department of Philosophy at the University of Alberta, and to the members of my family for their support and encouragement.

INTRODUCTION

Aesthetics and the environment

What is "environmental aesthetics"?

Aesthetics is the area of philosophy that concerns our appreciation of things as they affect our senses, and especially as they affect them in a pleasing way. As such it frequently focuses primarily on the fine arts, the products of which are traditionally designed to please our senses. However, much of our aesthetic appreciation is not confined to art, but directed toward the world at large. We appreciate not only art, but also nature – broad horizons, fiery sunsets, and towering mountains. Moreover, our appreciation reaches beyond pristine nature to our more mundane surroundings: the solitude of a neighborhood park on a rainy evening, the chaos of a bustling morning marketplace, the view from the road. Thus, there is a need for an aesthetics of the environment, for in such cases our aesthetic appreciation encompasses our surroundings: our environment. The environment may be more or less natural, large or small, mundane or exotic, but in each such case it is an environment that we appreciate. Such appreciation is the subject matter of environmental aesthetics.

The nature of environmental aesthetics

The fact that the focus of aesthetic appreciation is an environment signals several important dimensions of such appreciation which in turn determine the nature of environmental aesthetics. The first of these dimensions follows from the very fact that the object of appreciation, the "aesthetic object," is our environment, our surroundings. Thus, we as appreciators are immersed within the object of our appreciation. This fact has a number of ramifications: not only are we in what we appreciate but what we appreciate is also that from which we appreciate. If we move, we move within the object of our appreciation and thereby change our relationship to it and at the same time change the object itself. Moreover, since it is our surroundings, the object of appreciation impinges upon all our senses. As we occupy it or move through it, we see, hear, feel, smell, and perhaps even taste it. In short, the experience of the environmental object of appreciation from which aesthetic appreciation must be fashioned is initially intimate, total, and engulfing.

These dimensions of our experience are intensified by the unruly and chaotic nature of the object of appreciation itself. It is not the more or less discrete, stable, and self-contained object of traditional art, but rather an environment. Consequently, not only does it change as we move within it, it changes of its own accord. Environments are constantly in motion, in both the short and long term. If we remain motionless, the wind yet brushes our face and the clouds yet pass before our eyes. And with time changes continue without limit: night falls, days pass, seasons come and go. Moreover, environments not only move through time, they extend through space, and again without limit. There are no boundaries for our environment; as we move, it moves with us and changes, but does not end. Indeed, it continues unending in every direction. In other words, the environmental object of appreciation is not "framed" as are traditional works of art, neither in time as are dramatical works or musical compositions nor in space as are paintings or sculptures.

These differences between environments and traditional artistic objects relate to a deeper difference between the two. Works of art are the products of artists. The artist is quintessentially a designer, creating a work by embodying a design in an object. Thus, works of art are tied to their designers both causally and conceptually: what a work is and what it means follows from its designer and its design. However, environments typically are not the products of designers and typically have no design. Rather they come about "naturally," they change, grow, and develop by means of natural processes. Or they come about by means of human agency, but even then only rarely are they the result of a designer embodying a design. In short, the paradigm of the environmental object of appreciation is unruly in yet another way: neither its nature nor its meaning are determined by a designer and a design.

The upshot is that in our aesthetic appreciation of the world at large we are confronted by, if not intimately and totally engulfed in, something that forces itself upon all our senses, is constantly in motion, is limited neither in time nor in space, and is constrained concerning neither its nature nor its meaning. We are immersed in a potential object of appreciation and our task is to achieve aesthetic appreciation of that object. Moreover, the appreciation must be fashioned anew, with neither the aid of frames, the guidance of designs, nor the direction of designers. Thus, in our aesthetic appreciation of the world at large we must begin with the most basic questions, those of exactly what to aesthetically appreciate and how to appreciate it. These questions set the agenda for environmental aesthetics; the field essentially concerns the issue of what resources, if any, are available for answering them.

The two basic orientations in environmental aesthetics

The questions of what and how to aesthetically appreciate in an environment generate a number of different approaches, but at the most fundamental level two main points of view can be identified. The first may be characterized as subjectivist or perhaps as skeptical. In essence, it holds that since in the appreciation of environments we seemingly lack the resources normally involved in aesthetic appreciation, these questions cannot be properly answered. In other words, since we lack resources such as frames, designs, and designers as well as the guidance they provide, then we must embrace either subjectivism or skepticism: either there is no appropriate or correct aesthetic appreciation of environments or such appreciation as there is, is not real aesthetic appreciation. Concerning the world at large, as opposed to works of art, the closest we come to appropriate aesthetic appreciation is simply to open ourselves to being immersed, respond as we will, and enjoy what we can. And the question of whether or not the resultant experience is appropriate in some sense, or even aesthetic in any sense, is not of any importance.

A second basic point of view may be characterized as objectivist. In essence, it argues that, in addressing the what and how questions, there are in fact two resources to draw upon: the appreciator and the object of appreciation. Thus, roles that are played in the appreciation of traditional art objects by designer and design must be played in the aesthetic appreciation of an environment by either or both of these two resources. In such appreciation the role of designer is typically taken up by the appreciator and that of design by the object. In other words, in our aesthetic appreciation of the world at large we as appreciators typically play the role of artist and let the world provide us with something like a design. Thus, when confronted by an environment, we select the senses relevant to its appreciation and set the frames that limit it in time and space. Moreover, as designer plays off against design, so too in selecting and setting we play off against the nature of the environment we confront. In this way the environment by its own nature provides the analogue of a design – we might say it provides its own design. Thus, it offers the necessary guidance in light of which we, by our selecting and setting, can appropriately answer the questions of what and how to appreciate. We thereby fashion our initially engulfing if not overwhelming experience of an environment into appropriate and genuine aesthetic appreciation.

In disputes between subjectivist or skeptical positions and more objectivist ones, the burden of proof typically falls on the latter. To make its case, the objectivist account must be elaborated and supported by arguments and examples. The basic idea of the objectivist point of view is that our appreciation is guided by the nature of the object of appreciation. Thus, information about the object's nature, about its genesis, type, and properties, is necessary for appropriate aesthetic appreciation. For example, in appreciating a natural environment such as an alpine meadow, it is important to know, for instance, that it survives under constraints imposed by the climate of high altitude. With such knowledge comes the understanding that diminutive size in flora is an adaptation to such constraints. This knowledge and understanding guides our framing of the environment so that, for example, we avoid imposing inappropriately large frames, which may cause us simply to overlook miniature wild flowers. In such a case we might neither appreciatively note their remarkable adjustment to their situation nor attune our senses to their subtle fragrance, texture, and hue. Similarly, in appreciating human-altered environments such as those of modern agriculture, knowledge about, for example, the functional utility of cultivating huge fields devoted to single crops is aesthetically relevant. Such knowledge encourages us to enlarge and adjust our frames, our senses, and even our attitudes. As a result we may more appreciatively accommodate the vast uniform landscapes that are the inevitable result of such farming practices.

The scope of environmental aesthetics

Whether they endorse a subjectivist or an objectivist point of view, both basic orientations in environmental aesthetics recognize the array of special problems that confronts the field. Similarly, both recognize the expansive scope of the field itself. The scope may be characterized in terms of three continuums.

On the first, the subject matter of environmental aesthetics stretches from pristine nature to the very limits of the most traditional art forms, and by some accounts even expands to include the latter. On this continuum the things treated by environmental aesthetics range from wilderness areas, through rural landscapes and countrysides, to cityscapes, neighborhoods, market places, shopping centers, and beyond. Thus, within the genus of environmental aesthetics fall a number of different species, such as the aesthetics of nature, landscape aesthetics, the aesthetics of cityscapes and urban design, and perhaps the aesthetics of architecture, if not that of art itself.

The second continuum in terms of which the scope of environmental aesthetics may be characterized ranges over size. Many environments that are typical objects of our aesthetic appreciation, especially those that surround and threaten to engulf us, are very large: a dense old growth forest, a seemingly endless field of wheat, the downtown of a big city. But environmental aesthetics also focuses on smaller and more intimate environments, such as our backyard, our office, and our living room. And perhaps the scope extends even to diminutive environments, such as we may encounter when we turn over a rock or when traveling with a microscope into a drop of pond water. Such tiny environments, although not physically surrounding, are yet totally engaging.

The third continuum is closely related to the second. It ranges from the extraordinary to the ordinary, from the exotic to the mundane. Just as environmental aesthetics is not limited to the large, it is likewise not limited to the spectacular. Ordinary scenery, commonplace sights, and our day-to-day experiences are also proper objects of aesthetic appreciation. As such they not only fall under the scope of environmental aesthetics, but also, in light of becoming objects of aesthetic appreciation, hopefully become somewhat less ordinary.

In spite of the expansive scope of environmental aesthetics, a basic assumption of the field is that every environment, natural, rural, or urban, large or small, ordinary or extraordinary, offers much to see, to hear, to feel, much to aesthetically appreciate. In short, the different environments of the world at large are as aesthetically rich and rewarding as are works of art. Nonetheless, there are, as noted, special issues in aesthetic appreciation posed by the very nature of environments, by the fact that they are our surroundings, that they are unruly and chaotic objects of appreciation, and that we are plunged into them without appreciative guidelines. The chapters that follow embrace this basic assumption and recognize these special issues. They address the issues from an objectivist point of view.

Part I

THE APPRECIATION OF NATURE

1

THE AESTHETICS OF NATURE

A brief historical overview

In the Western world there has been since antiquity a tradition of viewing art as the mirror of nature. However, the idea of aesthetically appreciating nature itself is sometimes traced to a less ancient origin: Petrarch's novel passion for climbing mountains simply to enjoy the prospect. Yet even if the aesthetic appreciation of nature only dates from the dawn of the Renaissance, its development from that time to the present has been uneven and episodic. Initially, nature's appreciation as well as its philosophical investigation were hamstrung by religion. The reigning religious tradition could not but deem nature an unworthy object of aesthetic appreciation, for it saw mountains as despised heaps of wreckage left by the flood, wilderness regions as fearful places for punishment and repentance, and all of nature's workings as poor substitutes for the perfect harmony lost in humanity's fall. It took the rise of a secular science and equally secular art forms to free nature from such associations and thereby open it for aesthetic appreciation. Thus, in the Western world the evolution of aesthetic appreciation of nature has been intertwined with both the objectification of nature achieved by science and the subjectification of it rendered by art.

Although the scientific objectification of nature had earlier origins, the connection between aesthetic appreciation of nature and scientific objectivity dates from early in the eighteenth century. At that time, British aestheticians initiated a tradition that gave theoretical expression to this connection. Empiricist thinkers, such as Joseph Addison and Francis Hutcheson, took nature rather than art as the ideal object of aesthetic experience and developed the notion of disinterestedness as the mark of such experience. In the course of the century, this notion was elaborated such as to exclude from aesthetic experience an ever-increasing range of associations and conceptualizations. Thus, the objects of appreciation favored by this tradition, British landscapes, were, by means of disinterested aesthetic appreciation, eventually severed not only from religious associations, but from any appreciator's personal, moral, and economic interests. The upshot was a mode of aesthetic appreciation that looked upon the natural world with an eye not unlike the distancing, objectifying eye of science. In this way, the tradition laid the groundwork for the idea of the sublime. By means of the sublime even the most threatening of nature's manifestations, such as mountains and wilderness, could be distanced and appreciated, rather than simply feared and despised.

However, the notion of disinterestedness not only laid the groundwork for the sublime, it also cleared the ground for another, quite different idea, that of the picturesque. This idea secured the connection between aesthetic appreciation of nature and the subjective renderings of nature in art. The term "picturesque" literally means "picture-like" and indicates a mode of appreciation by which the natural world is divided into artistic scenes. Such scenes aim in subject matter or in composition at ideals dictated by the arts, especially poetry and landscape painting. Thus, while disinterestedness and the sublime stripped and objectified nature, the picturesque dressed it in a new set of subjective and romantic images: a rugged cliff with a ruined castle, a deep valley with an arched bridge, a barren outcropping with a crofter's cottage. Like disinterestedness and the sublime, the picturesque had its roots in the theories of the early eighteenth century aestheticians, such as Addison, who thought that what he called the "works of nature" were more appealing when they resembled works of art. However, picturesque appreciation did not culminate until later in the century when it was popularized primarily by William Gilpin and Uvedale Price. At that time, it became the reigning aesthetic ideal of English tourists who pursued picturesque scenery in the Lake District and the Scottish Highlands. Indeed, the picturesque remains the mode of aesthetic appreciation associated with the form of tourism that sees and appreciates the natural world primarily in light of renderings of nature typical of travel brochures, calendar photos, and picture postcards.

After the close of the eighteenth century, the picturesque lingered on as a popular mode of aesthetic appreciation of nature. However, the philosophical study of the aesthetics of nature, after the flowering of that century, went into steady decline. Many of the main ideas, such as the idea of the sublime, the notion of disinterestedness, and the theoretical centrality of nature rather than art, reached their climax with Kant. In his third critique some of these ideas received such exhaustive treatment that a kind of closure was seemingly achieved. Following Kant, a new world order was initiated by Hegel. In this world, art was a means to the Absolute, and it rather than nature was destined to became the favored subject of philosophical aesthetics.

However, even as the theoretical study of the aesthetics of nature declined, a new view of nature was initiated that eventually gave rise to a different kind of aesthetic appreciation. This mode of appreciation has its roots in the North American tradition of nature writing, as exemplified by Henry David Thoreau. In the middle of the nineteenth century, it was reinforced by the work of George Perkins Marsh and his recognition that humanity is the major cause of the destruction of nature's beauty. It achieved its classic realization at the end of the century with American naturalist John Muir. Muir saw all nature and especially wild nature as aesthetically beautiful and found ugliness only where nature was subject to human intrusion. These ideas strongly influenced the North American wilderness preservation movement and continue to shape the aesthetic appreciation of nature associated with contemporary environmentalism. This kind of appreciation may be called positive aesthetics.[1] In so far as positive aesthetic appreciation eschews humanity's marks on the natural landscape, it is somewhat the converse of picturesque appreciation with its delight in signs of human presence. Thus, it has become the rival of the picturesque as the popular mode of aesthetic appreciation of nature, although contemporary nature appreciation frequently involves a somewhat uneasy balance between the two different modes.

In spite of the developments in popular appreciation of nature in the nineteenth and twentieth century, however, philosophical aesthetics, with few exceptions, ignored nature throughout most of this period. In the nineteenth century Schelling and a scattering of thinkers of the Romantic Movement considered the aesthetics of nature to some extent, and in the first half of the twentieth century George Santayana and John Dewey each discussed it. But, by and large, in so far as aesthetics was pursued, it was completely dominated by an interest in art. Thus, by the mid-twentieth century, within the analytic tradition, philosophical aesthetics was virtually equated with philosophy of art. The major textbook in aesthetics at this time was subtitled *Problems in the Philosophy of Criticism* and major aesthetics anthologies bore titles such as *Art and Philosophy* and *Philosophy Looks at the Arts*.[2] Moreover, when aesthetic appreciation of nature was mentioned, it was treated, by comparison with that of art, as a messy, subjective business of little philosophical significance. However, in the second half of the twentieth century this situation was destined to change.

A brief overview of contemporary positions

Many of the issues in contemporary work on the aesthetics of nature are foreshadowed in one article: Ronald W. Hepburn's seminal "Contemporary Aesthetics and the Neglect of Natural Beauty."[3] After noting that by essentially reducing aesthetics to the philosophy of art, analytic aesthetics virtually ignores the natural world, Hepburn sets the agenda for the discussion of the late twentieth century. He argues that aesthetic appreciation of art frequently provides misleading guidelines for our appreciation of nature. Yet he observes that there is in the aesthetic appreciation of nature, as in appreciation of art, a distinction between appreciation which is only trivial and superficial and that which is serious and deep. He furthermore suggests that with nature such serious appreciation may require different approaches that can accommodate not only nature's indeterminate and varying character, but also both our multi-sensory experience and our diverse understanding of it.

The contemporary discussion of the aesthetics of nature thus stresses different approaches to or models for the appreciation of nature: models intended to capture the essence of appropriate aesthetic appreciation of nature. Certain more traditional models that are rather directly related to the aesthetic appreciation of the arts are seemingly inadequate. Two such models may be called the object model and the landscape model. The former pushes nature in the direction of sculpture and the latter treats it as similar to landscape painting. Thus, the object model focuses aesthetic appreciation primarily on natural objects and dictates appreciation of such objects rather as we might appreciate pieces of abstract sculpture, mentally or physically extracting them from their contexts and dwelling on their formal properties. On the other hand, the landscape model, following in the tradition of the picturesque noted in the first section of this chapter, mandates appreciation of nature as we might appreciate a landscape painting. This requires seeing it to some extent as a two-dimensional scene and again dwelling largely on formal properties. Neither of these models fully realize serious, appropriate appreciation of nature for each distorts the true character of nature. The former rips natural objects from their larger environments while the latter frames and flattens them into scenery. Moreover, in focusing mainly on formal properties, both models neglect much of our normal experience and understanding of nature.[4]

Although the aesthetic appreciation of the arts does not directly provide adequate models for the appreciation of nature, it yet suggests some of what is required in a more adequate model. In serious, appropriate aesthetic appreciation of works of art, it is essential that we appreciate works as what they in fact are and in light of knowledge of their real natures. Thus, for instance, serious, appropriate aesthetic appreciation of the *Guernica* (1937) requires that we appreciate it as a painting and moreover as a cubist or neo-cubist painting, and therefore that we appreciate it in light of our knowledge of paintings in general and of cubist paintings in particular. This suggests a third model for the aesthetic appreciation of nature, the natural environmental model. This model, which I develop throughout Part I of this volume, recommends two things. First, that, as in our appreciation of works of art, we must appreciate nature as what it in fact is, that is, as natural and as an environment. Second, it recommends that we must appreciate nature in light of our knowledge of what it is, that is, in light of knowledge provided by the natural sciences, especially the environmental sciences such as geology, biology, and ecology. The natural environmental model thus accommodates both the true character of nature and our normal experience and understanding of it.[5]

Nonetheless, the natural environmental model may be thought not to characterize our appropriate aesthetic appreciation of nature completely accurately. Although it does not, as the object and the landscape models, distort nature itself, it may yet be thought to somewhat misrepresent our appreciation of nature. Its emphasis on scientific knowledge gives such appreciation a highly cognitive and what may be judged an overly intellectual quality. In contrast to the cognitive emphasis of the natural environmental model, a fourth model, the engagement model, stresses the contextual dimensions of nature and our multi-sensory experience of it. Viewing the environment as a seamless unity of organisms, perceptions, and places, the engagement model beckons us to immerse ourselves in our natural environment in an attempt to obliterate traditional dichotomies such as subject and object, and ultimately to reduce to as small a degree as possible the distance between ourselves and nature. In short, aesthetic experience is taken to involve a total immersion of the appreciator in the object of appreciation.[6]

The engagement model calls for the absorption of the appreciator into the natural environment. Perhaps in doing so it goes too far. There are two main difficulties. First, in attempting to eliminate any distance between ourselves and nature, the engagement model may lose that by reason of which the resultant experience is aesthetic. As noted in the first section of this chapter, within the Western tradition the very notion of the aesthetic is conceptually tied to disinterestedness and the idea of distance between the appreciator and the appreciated. The second difficulty is that in attempting to obliterate dichotomies such as that between subject and object, the engagement model may also lose the possibility of distinguishing between trivial, superficial appreciation and that which is serious and appropriate. This is because serious, appropriate appreciation revolves around the object of appreciation and its real nature, while superficial appreciation frequently involves only whatever the subject happens to bring to the experience. In short, without the subject/object distinction, aesthetic appreciation of nature is in danger of degenerating into little more than a subjective flight of fancy.

Another view that also seems to diverge from the natural environmental model is the arousal model. This model challenges the central place the natural environmental model grants to scientific knowledge in aesthetic appreciation of nature. The arousal model holds that we may appreciate nature simply by opening ourselves to it and thus being emotionally aroused by it. The view contends that this less intellectual, more visceral experience of nature is a way of legitimately appreciating nature without involving any knowledge gained from science. Unlike the engagement model, this model does not call for a total immersion in nature, but only for an emotional relationship with it based on our common, everyday knowledge and experience of it. Consequently, in contrast to the engagement model, the arousal model does not lose the right to call its experience of nature aesthetic. Nor does it undercut the distinction between trivial and serious appreciation of nature, even though the appreciation it stresses may be more the former than the latter. However, the contrast between the arousal model and the natural environmental model is less clear. If we recognize our scientific knowledge of the natural world as only a finer-grained and theoretically richer version of our common, everyday knowledge of it, and not as something essentially different in kind, then the difference between the arousal model and the natural environmental model is mainly one of emphasis. Both models track the appreciation of nature, although the arousal model focuses on the more common, less cognitively rich, and perhaps less serious end of the continuum.[7]

A more fundamental challenge to the natural environmental model comes from what may be called the mystery model of nature appreciation. This view holds that the natural environmental model, in requiring that we must have knowledge of what we appreciate, has no place for the way in which nature is alien, aloof, distant, and unknowable. It contends that the only appropriate experience of nature is a sense of mystery involving a state of appreciative incomprehension, a sense of not belonging to and of being separate from nature. However, the mystery model faces major difficulties. With only mystery and aloofness, there seems to be no grounding for appreciation of any kind, let alone aesthetic appreciation. The mystery and aloofness of nature is a gulf, an emptiness, between us and nature; it is that by which we are separate from nature. Thus, mystery itself cannot constitute a means by which we can attain any appreciation of nature whatsoever. In short, insofar as nature is unknowable, it is also beyond aesthetic appreciation. However, even though mystery and aloofness cannot support appreciation, they can support worship. Thus, perhaps the mystery model should be characterized not as an aesthetic of nature, but rather as a religious approach to nature. If this is the case, then rather than revealing a dimension of our appropriate aesthetic appreciation of nature, the mystery model leaves the realm of the aesthetic altogether.[8]

If the mystery model of nature appreciation moves such appreciation outside the realm of the aesthetic, it does so unintentionally. However, the possibility that our appreciation of nature is not aesthetic is expressly embraced as the central tenet of the nonaesthetic model of nature appreciation. This view constitutes a radical alternative to all other models in explicitly claiming that nature appreciation is not a species of aesthetic appreciation. It holds that aesthetic appreciation is paradigmatically appreciation of works of art and is minimally appreciation of artifacts, of that which is human-made. Thus, in this view the appreciation of nature itself cannot be aesthetic appreciation of any kind whatsoever.[9] However, such a view is deeply problematic. The view finds some support in the tendency in analytic aesthetics noted in the first section of this chapter, that is, the tendency to reduce all of aesthetics to philosophy of art. Yet the view remains essentially counterintuitive. Many of our fundamental paradigms of aesthetic appreciation are instances of appreciation of nature, such as our appreciation of the radiance of a glowing sunset, the grace of a bird in flight, or the simple beauty of a flower. Moreover, the Western tradition in aesthetics, not to mention other traditions, such as the Japanese, is committed to a doctrine that explicitly excludes the nonaesthetic model of nature appreciation: the doctrine that, as one writer puts it, anything that can be viewed can be viewed aesthetically.[10]

There is nonetheless a grain of truth contained in the nonaesthetic model of nature appreciation that is worth preserving. The nonaesthetic model makes a virtue of what the mystery model encounters unintentionally: the fact that the more removed, the more separate, something is from humankind and its artifactualization, the more problematic is its aesthetic appreciation. The limiting case, inadvertently illuminated by the mystery model, is the complete impossibility of aesthetically appreciating the totally unknowable. The insight contained in the nonaesthetic model is thus that some degree of artifactualization is necessary for aesthetic appreciation. However, what this model fails to recognize is that our human conceptualization and understanding of nature is itself a minimal form of artifactualization. And although minimal, it is yet adequate to underwrite aesthetic appreciation. To aesthetically appreciate the natural world, we do not need to actually make it, as we make works of art; nor do we need to conceptualize it in artistic categories, as is done by the object and landscape models. When we cast the conceptual net of common-sense and scientific understanding over nature we do enough to it to make possible its aesthetic appreciation. This fact lends support to some other models, such as the natural environment model, for it suggests that granting a special place in nature appreciation to at least our common-sense and scientific knowledge of nature may be a necessary condition for providing an adequate account of serious, appropriate aesthetic appreciation of nature.

However, the realization that human conceptualization and understanding is a form of artifactualization adequate for underwriting genuine aesthetic appreciation of nature opens the door to further possibilities. The conceptual net of common-sense and scientific understanding is not the only one we cast over nature. There are also numerous other nets woven by human culture in its many forms – nets woven not only by art, but also by literature, folklore, religion, and myth. This realization suggests the possibility of what may be called a postmodern model of nature appreciation. Such a view would compare nature to a text, contending that in reading a text we appropriately appreciate not just the meaning its author intended, but any of various meanings that it may have acquired or that we may find in it. And, moreover, none of these possible meanings has priority; no reading of a text is privileged. Thus, on such a postmodern model, whatever cultural significance nature may have acquired and that we may find in it, the rich and varied deposits from our art, literature, folklore, religion, and myth, would all be accepted as proper dimensions of our aesthetic appreciation of nature. And of such dimensions none would be given priority; no particular appreciation would be privileged as more serious or more appropriate than any other.[11]

The possibility of a postmodern model of nature appreciation focuses attention on the many layers of human deposit that overlay pure nature. These layers range from the thin film of common sense, through the rich stratum of science, to the abundant accumulations of culture. In our encounters with nature we confront this diversity and a postmodern model would eagerly welcome it all. In sharp contrast, some other models, such as the engagement model and the mystery model, strive toward attempting to appreciate pure, unadulterated nature, to look on nature bare, as Euclid looked on beauty. But the nonaesthetic model demonstrates that to go too far in this direction is to go beyond the realm of the aesthetic, making any aesthetic appreciation of nature impossible.

Nonetheless, contained in the purist's inclination is the antidote for the potential excesses of a postmodern model. To achieve a balanced understanding of the situation, we must keep in mind that, as with appreciation of art, serious appreciation of nature means appreciating it as what it in fact is; and yet at the same time we must recognize that this also means appreciating nature as what it is *for us*. This idea limits yet enriches what is involved in appropriate aesthetic appreciation. Contra a postmodern model, it is not the case that just any fanciful reverie we happen to bring to nature will do as well as anything else; and contra the purist, nature is what it is for us, is what we have made of it. To miss or deny this latter fact is to miss or deny much of the richness that serious, appropriate aesthetic appreciation of nature has to offer.

The idea that nature is what it is for us, is what we have made of it, widens the scope of appropriate aesthetic appreciation, but it also constrains any view such as a postmodern model. In part this is because neither nature nor we are one unitary thing. It follows that not all of humankind's cultural deposit is aesthetically significant either to all parts of nature or for all of humankind. For any particular part of nature and for any particular appreciator, some of the cultural overlay is relevant, is indeed necessary for serious appreciation, while much of it is not, is indeed little better than fanciful daydreams. In light of this, perhaps what might be called a pluralist model of nature appreciation should supplant a postmodern model. A pluralist model would accept the diversity and the richness of the cultural overlay in which a postmodern model delights. However, such a model would also recognise, first, that for any particular part of nature only a small part of that cultural overlay is really relevant to serious, appropriate appreciation, and, second, that for any particular appreciator only a small part of the overlay can truly be claimed as his or her own. A pluralist model would endorse diversity, but yet would hold that in appropriate aesthetic appreciation, not all nature either can or should be all things to all human beings.[12]

Although a pluralist model would thus restrict the role of our cultural overlay in appropriate aesthetic appreciation of nature, any such restriction need not apply equally to all layers of the human deposit. Perhaps the more basic layers, those of common sense and science, are dimensions of appropriate appreciation of all of nature by any of its appreciators. Thus, even in light of the possibility of a pluralist model, models such as the arousal model and the natural environmental model maintain a special place as general guides to appropriate aesthetic appreciation of nature. This is because these models concentrate on the most fundamental layers of the human overlay, those constituting the very foundations of our experience and understanding of nature. However, there may be other layers relevant to our appreciation of nature that are equally universal, although they may constitute the spires rather than the foundations of the human deposit. Such layers are the focus of the metaphysical imagination model of nature appreciation. According to this view, our imagination interprets nature as revealing metaphysical insights: insights about the whole of experience, about the meaning of life, about the human condition, about humankind's place in the cosmos. Thus, this model includes in appropriate aesthetic appreciation of nature those abstract meditations and speculations about the true nature of reality that our encounters with nature frequently engender in us.[13]

The metaphysical imagination model invites us to entertain in our aesthetic appreciation of nature deep meditations and possibly wild speculations, but again the question of what is and what is not relevant arises. Which of such meditations and speculations are only trivial and fanciful and which are serious and sustainable? In essence, the question is again that of what does and what does not actually focus on and reveal nature as it in fact is. However, in the context of the metaphysical imagination model, this question arises as a general and profound question about the real nature of the natural world. Thus, the metaphysical imagination model indicates a new agenda for the aesthetics of nature, for it suggests that in order to ultimately adjudicate among the different models of aesthetic appreciation of nature, we must first resolve more fundamental metaphysical issues about the true character of nature and about our proper place in its grand design.[14]

The new agenda for the aesthetics of nature indicated by the metaphysical imagination model points to the need to address fundamental issues about the nature of the natural world and our place in it. Thus, this agenda suggests that primary consideration should be given to those models of aesthetic appreciation of nature that most directly deal with such issues. As noted in the first section of this chapter, within the Western world the development of the aesthetic appreciation of nature has been closely intertwined with the growth of the natural sciences. And, of course, within the Western world, it is science that is taken to most successfully address fundamental issues about the true character of the natural world and humanity's place in it. Consequently, the new agenda points to the centrality of that model which ties appropriate aesthetic appreciation of nature most closely to scientific knowledge: the natural environmental model.[15]

The natural environmental model: some further ramifications

In addition to directly speaking to the new agenda for the aesthetics of nature, the natural environmental model has a number of other ramifications worth noting. Some of these concern what is called applied aesthetics, in particular popular appreciation of nature as practiced not only by tourists but by each of us in our daily pursuits. As noted in the first section of this chapter, such appreciation frequently involves a somewhat uneasy balance between two different modes: on the one hand, that flowing from the tradition of the picturesque and currently embodied in artistic and cultural models, such as the landscape model, and, on the other, the positive aesthetics mode growing from the tradition of thinkers such as Thoreau and Marsh and fully realized in John Muir. The balance between the two modes of appreciation is to some degree tipped in favor of the latter by the natural environmental model in that this model provides theoretical underpinnings for positive aesthetics. When nature is aesthetically appreciated in virtue of the natural and environmental sciences, positive aesthetic appreciation is singularly appropriate, for, on the one hand, pristine nature – nature in its natural state – is an aesthetic ideal and, on the other, as science increasingly finds, or at least appears to find, unity, order, and harmony in nature, nature itself, when appreciated in light of such knowledge, appears more fully beautiful.[16]

Other ramifications of the natural environmental model are more directly environmental and ethical. Many of the other models for the aesthetic appreciation of nature are frequently condemned as totally anthropocentric, as not only anti-natural but also arrogantly disdainful of environments that do not conform to artistic and cultural ideals and preconceptions. The root source of these environmental and ethical concerns is that such models, as noted in the second section of this chapter, do not always encourage appreciation of nature for what it is and for the qualities it has. However, since the natural environmental model bases aesthetic appreciation on a scientific view of nature, it thereby endows aesthetic appreciation of nature with a degree of objectivity that helps to dispel environmental and moral criticisms, such as the charge of anthropocentrism. Moreover, the possibility of an objective basis for aesthetic appreciation of nature also holds out promise of more direct practical relevance in a world increasingly engaged in environmental assessment.[17] Individuals making such assessments, although typically not worried about anthropocentrism, are yet frequently reluctant to acknowledge the relevance and importance of aesthetic considerations, regarding them simply as at worst completely subjective whims or at best only relativistic, transient, and soft-headed cultural ideals. Recognizing that aesthetic appreciation of nature has scientific underpinnings helps to meet such doubts.

Another consequence concerns the discipline of aesthetics itself. The natural environmental model, in rejecting artistic and other related models of nature appreciation in favour of a dependence on common sense/scientific knowledge, provides a blueprint for aesthetic appreciation in general. This model suggests that in aesthetic appreciation of anything, be it people or pets, farmyards or neighborhoods, shoes or shopping malls, appreciation must be centered on and driven by the real nature of the object of appreciation itself.[18] In all such cases, what is appropriate is not an imposition of artistic or other inappropriate ideals, but rather dependence on and guidance by means of knowledge, scientific or otherwise, that is relevant given the nature of the thing in question.[19] This turn away from irrelevant preconceptions and toward the real nature of objects of appreciation points the way to a general aesthetics that expands the traditional conception of the discipline, which, as noted in the first section of this chapter, has been for much of this century narrowly equated with the philosophy of art. The upshot is a more universal aesthetics. This is the field of study, now generally termed environmental aesthetics, that is delineated in the general introduction to this volume. Its relevance to the natural world is investigated in Part I of the volume; its application beyond the realm of nature is pursued in Part II.

Lastly, in initiating a more universal and object-centered environmental aesthetics, the natural environmental model aids in the alignment of aesthetics with other areas of philosophy, such as ethics, epistemology, and philosophy of mind, in which there is increasingly a rejection of archaic, inappropriate models and a new-found dependence on knowledge relevant to the particular phenomena in question. For example, the natural environmental model, in its rejection of appreciative models condemned as anthropocentric, parallels environmental ethics in the latter's rejection of anthropocentric models for the moral assessment of the natural world and the replacement of such models with paradigms drawn from the environmental and natural sciences. The general challenge is that we confront a natural world that allows great liberty concerning the ways and means of approaching it, and that we must therefore find the right models in order to treat it appropriately. The aesthetic dimension of this challenge is the primary focus of the remaining chapters of the first part of this volume.

Notes

1 I discuss positive aesthetics in more detail in "Nature and Positive Aesthetics," *Environmental Ethics*, 1984, vol. 6, pp. 5–34 (reproduced in this volume, Chapter 6). Muir's view is well exemplified in his *Atlantic Monthly* essays collected in *Our National Parks*, New York, Houghton Mifflin, 1916. For an introduction to the nature of picturesque appreciation, see M. Andrews, *The Search for the Picturesque*, Stanford, Stanford University Press, 1989.

2 The volumes referred to here are Monroe C. Beardsley's important text, *Aesthetics: Problems in the Philosophy of Criticism*, New York, Harcourt, Brace & World, 1958, and the first two major anthologies in analytic aesthetics, W. E. Kennick (ed.) *Art and Philosophy: Readings in Aesthetics*, New York, St Martin's Press, 1964 and Joseph Margolis (ed.) *Philosophy Looks at the Arts: Contemporary Readings in Aesthetics*, New York, Charles Scribner's Sons, 1962. It is remarkable that even with a total of 1,527 pages among them, none of these volumes, each a classic of its kind, so much as mentions the aesthetics of nature.

3 Ronald W. Hepburn, "Contemporary Aesthetics and the Neglect of Natural Beauty," in Bernard Williams and Alan Montefiore (eds) *British Analytical Philosophy*, London, Routledge and Kegan Paul, 1966, pp. 285–310. Ronald W. Hepburn, "Aesthetic Appreciation of Nature," in Harold Osborne (ed.) *Aesthetics in the Modern World*, London, Thames and Hudson, 1968, pp. 49–66, is a shorter version of the same article. Also see Ronald W. Hepburn, "Trivial and Serious in Aesthetic Appreciation of Nature," in Salim Kemal and Ivan Gaskell (eds) *Landscape, Natural Beauty, and the Arts*, Cambridge, Cambridge University Press, 1993, pp. 65–80.

4 I critique the object and the landscape models in more detail in "Appreciation and the Natural Environment," *Journal of Aesthetics and Art Criticism*, 1979, vol. 37, pp. 267–76 (reproduced in this volume, Chapter 4). For an updated version of this essay,

see "Aesthetic Appreciation and the Natural Environment," in S. Armstrong and R. Botzler (eds) *Environmental Ethics: Divergence and Convergence*, Second Edition, New York, McGraw Hill, 1998, pp. 122–31 or, for a shorter version, see the same title in S. Feagin and P. Maynard (eds) *Aesthetics*, Oxford, Oxford University Press, 1997, pp. 30–40.

5 The initial development of the natural environment model is in "Appreciation and the Natural Environment," op. cit., (reproduced in this volume, Chapter 4) and in "Nature, Aesthetic Judgment, and Objectivity," *Journal of Aesthetics and Art Criticism*, 1981, vol. 40, pp. 15–27 (reproduced in this volume, Chapter 5).

6 The original and best development of the engagement model is in Arnold Berleant, *The Aesthetics of Environment*, Philadelphia, Temple University Press, 1992. See also Berleant's *Art and Engagement*, Philadelphia, Temple University Press, 1991 and *Living in the Landscape: Toward an Aesthetics of Environment*, Lawrence, University Press of Kansas, 1997. I discuss some of the philosophical underpinnings of the engagement model in "Beyond the Aesthetic," *Journal of Aesthetics and Art Criticism*, 1994, vol. 52, pp. 239–41 and "Aesthetics and Engagement," *British Journal of Aesthetics*, 1993, vol. 33, pp. 220–27.

7 The arousal model is presented in Noel Carroll, "On Being Moved By Nature: Between Religion and Natural History," in Kemal and Gaskell, opt. cit., pp. 244–66. I discuss this model in detail and develop the suggestions made here in "Nature, Aesthetic Appreciation, and Knowledge," *Journal of Aesthetics and Art Criticism*, 1995, vol. 53, pp. 393–400.

8 The mystery model is found in Stan Godlovitch "Icebreakers: Environmentalism and Natural Aesthetics," *Journal of Applied Philosophy*, 1994, vol. 11, pp. 15–30. I discuss the mystery model more fully in "Appreciating Godlovitch," *Journal of Aesthetics and Art Criticism*, 1997, vol. 55, pp. 55–7 and "Nature, Aesthetic Appreciation, and Knowledge," op. cit.

9 The nonaesthetic model, under the name the Human Chauvinistic Aesthetic, is outlined in Don Mannison, "A Prolegomenon to a Human Chauvinistic Aesthetic," in Don Mannison, Michael McRobbie, and Richard Routley (eds) *Environmental Philosophy*, Canberra, Australian National University, 1980, pp. 212–16. A similar position is defended in Robert Elliot, "Faking Nature," *Inquiry*, 1982, vol. 25, pp. 81–93. But for a revised position see Robert Elliot, *Faking Nature: The Ethics of Environmental Restoration*, London, Routledge, 1997. I discuss this kind of approach in "Nature and Positive Aesthetics," op. cit. (reproduced in this volume, Chapter 6) and "Appreciating Art and Appreciating Nature," in Kemal and Gaskell, op. cit., pp. 199–227 (reproduced in this volume, Chapter 7).

10 See Paul Ziff, "Anything Viewed," in *Antiaesthetics: An Appreciation of the Cow with the Subtile Nose*, Dordrecht, Reidel, 1984, pp. 129–39.

11 Although the postmodern model seems a recent innovation, it was suggested over one hundred years ago by George Santayana in *The Sense of Beauty*, [1896], New York, Collier, 1961, p. 99. He claims that the natural landscape is "indeterminate" and must be "composed" by each of us by being "poetized by our day-dreams, and turned by our instant fancy into so many hints of a fairyland of happy living and vague adventure" before "we feel that the landscape is beautiful." I discuss the postmodern model in "Between Nature and Art" (in this volume, Chapter 8) and in "Landscape and Literature" (in this volume, Chapter 14). I consider postmodern architecture in "Existence, Location, and Function: The Appreciation of Architecture," in M. Mitias (ed.) *Philosophy and Architecture*, Amsterdam, Rodopi, 1994, pp. 141–64 (reproduced in this volume, Chapter 13).

12 Some themes of the pluralist model have been suggested to me by the work of Yrjo Sepanmaa, especially his *The Beauty of Environment*, Second Edition, Denton, Texas,

Environmental Ethics Books, 1993. I endorse certain aspects of this model in "Landscape and Literature" (in this volume, Chapter 14).

13 The metaphysical imagination model is outlined in Ronald W. Hepburn, "Landscape and the Metaphysical Imagination," *Environmental Values*, 1996, vol. 5, pp. 191–204.

14 Since, as noted, Hepburn's "Contemporary Aesthetics and the Neglect of Natural Beauty," op. cit., set the initial agenda for the contemporary discussion of the aesthetics of nature and since the metaphysical imagination model is one of his most recent contributions to the field, it is most fitting that this model should set a new agenda for future discussion.

15 Perhaps a paradigm exemplification of aesthetic appreciation enhanced by natural science is Aldo Leopold's *A Sand County Almanac*, Oxford, Oxford University Press, 1949. Leopold's aesthetics are elaborated in "The Land Aesthetic," in J. B. Callicott (ed.) *Companion to A Sand County Almanac*, Madison, University of Wisconsin Press, 1987, pp. 157–71. The centrality of science in aesthetic appreciation of nature is challenged in Y. Saito, "Is There a Correct Aesthetic Appreciation of Nature?," *Journal of Aesthetic Education*, 1984, vol. 18, pp. 35–46. I discuss her concerns in "Saito on the Correct Aesthetic Appreciation of Nature," *Journal of Aesthetic Education*, 1986, vol. 20, pp. 85–93. For further discussion of the role of science, see H. Rolston, "Does Aesthetic Appreciation of Nature Need to be Science Based?," *British Journal of Aesthetics*, 1995, vol. 35, pp. 374–86 and M. Budd, "The Aesthetic Appreciation of Nature," *British Journal of Aesthetics*, 1996, vol. 36, pp. 207–22, as well as Y. Saito, "The Aesthetics of Unscenic Nature," S. Godlovitch, "Evaluating Nature Aesthetically," C. Foster, "The Narrative and the Ambient in Environmental Aesthetics," E. Brady, "Imagination and the Aesthetic Appreciation of Nature," M. M. Eaton, "Fact and Fiction in the Aesthetic Appreciation of Nature," and H. Rolston, "Aesthetic Experience in Forests," all in A. Berleant and A. Carlson (eds) *The Journal of Aesthetics and Art Criticism: Special Issue: Environmental Aesthetics*, 1998, vol. 56, pp. 101–66.

16 I consider positive aesthetics and its relationship to the rise of science in "Nature and Positive Aesthetics," op. cit, (reproduced in this volume, Chapter 6).

17 I discuss objectivity in appreciation of nature in "Nature, Aesthetic Judgment, and Objectivity," op. cit. (reproduced in this volume, Chapter 5) and in "On the Possibility of Quantifying Scenic Beauty," *Landscape Planning*, 1977, vol. 4, pp. 131–72. I briefly overview some of the issues in landscape assessment research in "Landscape Assessment," in M. Kelly (ed.) *Encyclopedia of Aesthetics*, New York, Oxford University Press, 1989, Vol. 3, pp. 102–5.

18 I apply these ideas to other kinds of cases in Part II of this volume, especially in "On the Aesthetic Appreciation of Japanese Gardens," *British Journal of Aesthetics*, 1997, vol. 37, pp. 47–56 (reproduced in this volume, Chapter 11), "On Appreciating Agricultural Landscapes," *Journal of Aesthetics and Art Criticism*, 1985, vol. 43, pp. 301–12 (reproduced in this volume, Chapter 12), and "Existence, Location, and Function: The Appreciation of Architecture," in Mitias, op. cit. (reproduced in this volume, Chapter 13).

19 I develop the idea of an object-centered aesthetics in "Appreciating Art and Appreciating Nature," in Kemal and Gaskell, op. cit., pp. 199–227 (reproduced in this volume, Chapter 7) and throughout Part II of this volume, especially in "Between Nature and Art" (this volume, Chapter 8).

2

UNDERSTANDING AND AESTHETIC EXPERIENCE

Aesthetic experience on the Mississippi

In *Life on the Mississippi* Mark Twain remarks:

> The face of the water, in time, became a wonderful book – a book that was a dead language to the uneducated passenger, but which told its mind to me without reserve, delivering its most cherished secrets as clearly as if it uttered them with a voice…In truth, the passenger who could not read this book saw nothing but all manner of pretty pictures in it, painted by the sun and shaded by the clouds, whereas to the trained eye these were not pictures at all, but the grimmest and most dead-earnest of reading matter.
>
> Now when I had mastered the language of this water…I had made a valuable acquisition. But I had lost something, too. I had lost something which could never be restored to me while I lived. All the grace, the beauty, the poetry had gone out of the majestic river! I still kept in mind a certain wonderful sunset which I witnessed when steamboating was new to me. A broad expanse of the river was turned to blood; in the middle distance the red hue brightened into gold, through which a solitary log came floating, black and conspicuous; in one place a long, slanting mark lay sparkling upon the water; in another the surface was broken by boiling, tumbling rings, that were as many-tinted as an opal; where the ruddy flush was faintest, was a smooth spot that was covered with graceful circles and radiating lines, ever so delicately traced; the shore on our left was densely wooded, and the somber shadow that fell from this forest was broken in one place by a long, ruffled trail that shone like silver; and high above the forest wall a clean-stemmed dead tree waved a single leafy bough that glowed like a flame in the unobstructed splendor that was flowing from the sun. There were graceful curves, reflected images, woody heights, soft distances; and over the whole scene, far and near, the dissolving lights drifted steadily, enriching it, every passing moment, with new marvels of coloring.

> I stood like one bewitched. I drank it in, in a speechless rapture...But as I have said, a day came when...if that sunset scene had been repeated, I should have looked upon it without rapture, and should have commented upon it, inwardly, after this fashion: The sun means that we are going to have wind to-morrow; the floating log means that the river is rising, small thanks to it; that slanting mark on the water refers to a bluff reef which is going to kill somebody's steamboat one of these nights, if it keeps on stretching out like that; those tumbling "boils" show a dissolving bar and a changing channel there; the lines and circles in the slick water over yonder are a warning that that troublesome place is shoaling up dangerously; that silver streak in the shadow of the forest is the "break" from a new snag, and he has located himself in the very best place he could have found to fish for steamboats; that tall dead tree, with a single living branch, is not going to last long, and then how is a body ever going to get through this blind place at night without the friendly old landmark?
>
> No, the romance and the beauty were all gone from the river.[1]

In this passage, Twain describes two different experiences of the river. The first, when steamboating was new to him and he saw the river as would an "uneducated passenger," is an experience of "all manner of pretty pictures, painted by the sun and shaded by the clouds." The second, when he had "mastered the language of this water" and he looked upon the river with a "trained eye," is an experience in which the pretty pictures are replaced by an understanding of the meaning of the river. Twain suggests that the two experiences are mutually exclusive, or at least that each makes the other difficult, if not impossible. On the one hand, the uneducated passenger sees "nothing but" pretty pictures because he or she cannot read the language of the river. On the other, Twain, once he had learned to understand this language, had "lost something which could never be restored" – "All the grace, the beauty, the poetry had gone out of the majestic river!"

The first experience Twain describes, that of the grace, the beauty, and the poetry of the river, is of the kind that is typically called aesthetic experience. The second is of the kind that may be characterized as cognitive; it involves an understanding of meanings achieved in virtue of knowledge gained through education or training. Thus, a significant question posed by Twain's remarks is the question of whether or not aesthetic experience and cognitive experience are in conflict in the way in which he seemingly suggests that they are. Is it the case that, without knowledge and understanding of that which we experience, we, like the uneducated passenger, may experience it aesthetically? And, more important, is it the case that once we have acquired such knowledge and understanding the possibility of aesthetic experience is in some way destroyed – and that we, like Twain, have then "lost something which could never be restored"?

To address these issues, we need to examine more carefully Twain's conception of the nature of aesthetic experience. Twain says that without knowing the language of the river, the uneducated passenger sees nothing but pretty pictures. However, his account of his own aesthetic experience of the river before he had acquired this knowledge of the river's language reveals an experience more cultivated than simply seeing pretty pictures. It is an experience of overpowering beauty which "bewitches" him, reducing him to a "speechless rapture." Moreover, his description of the scene that evokes this rapture is primarily in terms of two kinds of things. First, he describes "marvels of coloring" such as "the river…turned to blood,…the red hue brightened into gold," "rings…as many-tinted as an opal," a "trail that shone like silver," and a "bough that glowed like a flame in the unobstructed splendor that was flowing from the sun." Second, he describes what may be called marvels of form, such as "a long, slanting mark…sparkling upon the water," "the surface broken by boiling, tumbling rings," and "a smooth spot…covered with graceful circles and radiating lines, ever so delicately traced." In short, Twain's conception of aesthetic experience is that of an intense emotional state evoked by the appreciation of striking combinations of colors, lines, and shapes. Twain's experience is both more refined and in a sense sparser than that of seeing pretty pictures.

Formalism and aesthetic experience

If this is a correct characterization of Twain's conception of aesthetic experience, we may now inquire as to why, given this conception, he seemingly holds that an aesthetic experience of a thing is in conflict with knowledge yielding an understanding of its meaning. However, such an inquiry is hampered by the fact that other than a few scattered remarks such as those quoted here, Twain does not have much to say about aesthetic matters. Consequently, perhaps this inquiry can be most fruitfully pursued by considering Twain's views in light of a familiar theory of art – a theory to which his views are remarkably similar. The theory is what is known as the formalist theory of art; it is associated with the so-called "art for art's sake movement" which was fashionable around the turn of the century. The theory receives its best-known treatment in the writings of two British art critics, Clive Bell and Roger Fry. Bell's development of the theory is especially helpful in attempting to understand Twain's point of view.

In the tradition of classic philosophy of art, Bell develops the formalist theory as a theory providing a definition of art in terms of an essential quality, a quality, as he puts it, "common and peculiar to all members" of the class of works of art.[2] This essential quality he calls "significant form" and characterizes as aesthetically moving "relations and combinations of lines and colors."[3] Moreover, he holds that such moving combinations of lines and colors evoke in "anyone capable of feeling it" a particular kind of emotion which he terms the "aesthetic emotion."[4] Thus, Bell's account of aesthetic experience is essentially like that of Twain in that the experience is constituted by a heightened emotional state evoked by the formal aspects of objects. Just as Twain speaks of his experience of being moved to "speechless rapture" by the forms and colors of the river, Bell speaks of his "thrilling raptures," saying of himself that "in those moments I lose myself in that infinitely sublime state of mind to which pure visual form transports me."[5] And although Bell takes the experience of art to be the paradigm of this kind of formalist aesthetic experience, he also, like Twain, conceives of the aesthetic appreciation of nature in similar terms. He claims that to achieve "a thrill indistinguishable from that which art gives" an appreciator must see the landscape not "as fields and cottages" but rather contrive "to see it as a pure formal combination of lines and colours."[6]

Bell's account of the aesthetic experience suggests at least one way of understanding why Twain might believe that aesthetic experience is in conflict with knowledge yielding an understanding of meaning. Bell sharply distinguishes that which is essential to art, the significant form of a work, from its content. Moreover, he holds that while significant form evokes aesthetic emotion, content evokes only what he calls the emotions of life, our common emotional states unrelated to aesthetic emotion. As he puts it: "...he who contemplates a work of art inhabits a world with an intense and peculiar significance of its own; that significance is unrelated to the significance of life. In this world the emotions of life find no place."[7] However, for Bell the content of a work of art is its representational content which, in evoking the emotions of life, can detract from the intense, peculiar, and unrelated aesthetic emotion. Therefore, the content of a work is both irrelevant to and potentially destructive of aesthetic appreciation. In a well-known passage Bell declares: "The representative element in a work of art may or may not be harmful; always it is irrelevant."[8] And since the representational content of a work is, in one sense, the meaning of the work, the aesthetic appreciation of a work is thereby necessarily severed from any understanding of that meaning and the knowledge required to achieve such understanding. The upshot is that, in Bell's words, "to appreciate a work of art we need bring with us nothing from life, no knowledge of its ideas and affairs, no familiarity with its emotions. Art transports us from the world of man's activity to a world of aesthetic exaltation. For a moment we are shut off from human interests; our anticipations and memories are arrested; we are lifted above the stream of life."[9]

Thus, in formalism we have a theory of art which provides the following account of the conflict Twain finds between aesthetic experience and understanding: aesthetic experience involves only the appreciation of form and excludes any attention to content; the former requires no knowledge yielding understanding of meaning and, although the latter does require such knowledge, since it is excluded from aesthetic appreciation, whatever knowledge is needed to understand the meaning of content is at best irrelevant and at worse harmful to aesthetic appreciation. On this position aesthetic appreciation is a sparse, pure experience stripped of any associations with representational content and of any meanings it embodies. However, in order to maintain this narrow conception of aesthetic experience and to thereby account for Twain's conflict, this position requires that the distinction it draws between form and content be sharp, unequivocal, and impermeable. Only if the distinction is such, does it effectively rule out knowledge yielding understanding of meaning and put such understanding in essential opposition to pure aesthetic appreciation. Thus, it is necessary to examine this distinction more closely in order to determine whether or not it is viable and strong enough to account for Twain's conflict.

To investigate the form/content distinction, let us consider an example that, in spite of its representational content, is yet highly formal and should therefore put the distinction in its best light. Consider a work such as Aubrey Beardsley's *The Peacock Skirt* (1894), one of his well-known illustrations for Oscar Wilde's *Salome*. The work is a line drawing in back and white and depicts two individuals facing one another, one of whom is wearing a large, flowing skirt resembling a peacock's spread tail feathers. If we briefly describe each of the form and the content of this work on three different levels – that of the basic elements, the relationships among these elements, and the more general overall quality of the work – then *The Peacock Skirt* yields a form/content distinction something along the following lines. On the content side of the ledger we have representational content primarily consisting of: first, representations of two human beings, one of the human beings wearing a skirt resembling a peacock's tail; second, the two human beings represented as facing one another and talking, perhaps whispering, to one another; and, third, a general quality of refinement and elegance, but perhaps with a suggestion of something sinister or even evil. On the form side of the ledger we have primarily the following: first, as basic elements, two major shapes, one larger and rather pear-shaped and a second roughly rectangular, the former essentially in outline and the latter with a lined pattern throughout, and each consisting of black-and-white color and numerous lines; second, as relationships of these elements, the former shape is dominant and to the left of the composition and the latter recessive and to the right, while the white color is the background color with the black largely concentrated at the bottom of the composition and to a lesser extent at the top and along the right side; third, as general quality, a composition that is sparse, yet graceful, unified, and balanced.

Given this form/content distinction, the formalist theory holds that the aesthetic appreciation of *The Peacock Skirt* consists of only that which is on the form side of our ledger and that that which is on the content side is irrelevant and perhaps even harmful to such appreciation. The aesthetic experience of *The Peacock Skirt* is limited to the appreciation of the two major shapes together with the lines and colors that comprise them, the intricate relationships among these shapes, lines, and colors, and the grace, unity, and balance to which these elements and relationships give rise. But, we must ask, is such pure aesthetic appreciation possible and is the notion of such appreciation even

Illustration 1 *Peacock Skirt*, by Aubrey Beardsley (1894).

intelligible? Part of the problem is that this kind of appreciation seems at least psychologically very difficult, if not impossible. Exactly how are we to manage such appreciation? Psychologically speaking, how do we draw this sharp, unequivocal, and impermeable form/content distinction? How can we even see the shape of a human being without also seeing that it represents a human being and thereby understanding its meaning. Such appreciating of form without content can be facilitated by manoeuvres such as the following: turning the work upside down, moving far enough away from it such that its representations blur into abstract forms, squinting one's eyes so as to similarly blur the images, taking off one's glasses, if one wears glasses, or even just getting very tired or drunk. But is there not something wrong with a position that requires that we put either the work or ourselves in some kind of altered state in order to achieve aesthetic appreciation? Seemingly, the notion of aesthetic experience embodied in such a position is bordering on unintelligibility and incoherence. However, if confronted with such worries, Bell and Twain could simply, as they say, bite the bullet, holding that no one ever said that aesthetic appreciation was easy and, moreover, that the very difficulty of appreciating form without content even supports their own position, for it makes clear exactly how content can be harmful and how the understanding of meanings can destroy the possibility of aesthetic experience.

However, there are other difficulties with the formalist form/content distinction that are less easily put aside. Look again at *The Peacock Skirt* and consider its formal qualities such as its unified and balanced composition. Now imagine attempting to aesthetically appreciate it as the formalist theory of art requires, removing your glasses or squinting your eyes if necessary. However, to the extent that you achieve formalist aesthetic appreciation of the work, the work itself will appear less unified and balanced. Seemingly the unity evaporates in part because, as formalist aesthetic appreciation is achieved, the white center of the larger shape becomes a part of the background rather than a figure, and this sets free both the black patches at the top of the work and the large concentration of black at the bottom. This in turn undercuts the balance of the work by unstabilizing the relationships between the two major shapes and between the different areas of black, making the whole composition bottom-heavy. However, as long as aesthetic appreciation includes understanding of content, the unity and the balance of the composition are maintained, seemingly because the very fact that two human beings are represented stabilizes the two major shapes and moreover gives a powerful focal point in the human faces at the top of the work, which focal point is sufficient to balance the heavy black peacock skirt at the bottom. In a similar way, in Christian works of art balance is sometimes achieved between many large shapes and one small cross shape, not by means of formal elements and relationships, but rather by means of the meanings that the shapes possess. In short, in works such as *The Peacock Skirt* the overall formal quality of the work of art is dependent not only on formal elements and relationships but on content elements and relationships as well. Apparently the formalist's putatively sharp, unequivocal, and impermeable distinction between form and content has some cracks.

Moreover, it is not only the overall formal quality of a work that frequently depends on content. As our consideration of *The Peacock Skirt* suggests, without appreciation of representational content, the larger pear-like shape begins to disintegrate. However, the situation is in fact more serious than this. Without appreciation of the two major shapes of the work as representations of two human beings, seemingly there are no grounds for even saying that the work has two major shapes, rather than four or five or seven or seven hundred shapes. Imagine counting the shapes in *The Peacock Skirt* or in any representational work of art without reference to its content. This is like attempting to count the things in your study – without having a means of identifying what constitutes a thing. It seems that the very identity of the formal elements of a work of art such as *The Peacock Skirt* depends essentially upon the content of the work. The former cannot even be identified, let alone aesthetically appreciated, without reference to the latter. Thus, the formalist's sharp, unequivocal, and impermeable distinction between form and content is shown to be not simply cracked, but thoroughly porous, if it exists at all. In short, the form/content distinction simply cannot be successfully maintained in the way in which the formalist theory of art attempts to draw it; and the lean, formalist conception of absolutely pure aesthetic experience that depends upon this distinction is indeed exposed as incoherent.

Where does all this leave our attempt to understand Twain's observations concerning his two experiences of the river and his suggestion that there is a conflict between them? We have found that the formalist way of drawing the form/content distinction leads to incoherence. Therefore, the attempt, by means of this distinction, to exclude all knowledge facilitating the understanding of meaning from aesthetic experience is shown to be ungrounded. Even with the narrow formalist conception of aesthetic experience, which Bell and Twain share, no necessary conflict between aesthetic experience and understanding is established. In fact, the contrary seems to be the case: a certain level and kind of knowledge facilitating the understanding of meaning is apparently necessary in order to make any kind of aesthetic appreciation possible – even sparse formalist aesthetic appreciation. But in light of these findings we must ask: if Twain's narrow conception of aesthetic experience does not after all provide the resources for an account of the conflict he finds between aesthetic experience and understanding, then what, if anything, will?

Disinterestedness and aesthetic experience

To find the answer to this question, perhaps we need to consider the other dimension of Twain's observations, that is, the second of the two experiences he describes – not his aesthetic experience of the river, but the experience he had after he had "mastered the language of this water" and looked upon it with a "trained eye" – the experience in which moving combinations of colors, lines, and shapes are replaced by understanding of the meaning of the river. Concerning this second kind of experience it is important to note the nature of the meanings involved. In this experience of the river, meaning pertains to matters such as rising water, hidden reefs, changing channels, dangerous shoals, new snags, and blind spots; in short, the meaning the riverboat pilot must understand in order to prevent the river from "killing his steamboat." Twain notes that the language of the river he has learned is "never one that you could leave unread without loss," adding that the "passenger who could not read it was charmed with a peculiar sort of faint dimple on [the water's] surface...; but to the pilot that was an *italicized* passage;...for it meant that a wreck or a rock was buried there that could tear the life out of the strongest vessel that ever floated. It is the faintest and simplest expression that water ever makes, and the most hideous to a pilot's eye."[10]

In the way in which considering the formalist theory of art facilitated our examination of Twain's conception of aesthetic experience, considering another important tradition in aesthetic theory may help clarify the significance of what we may now call a "steamboat pilot experience" of the river. This is the tradition which attempts to analyze aesthetic experience in terms of a distinct kind of mental state typically described as one of disinterestedness. A significant contemporary advocate of disinterestedness as an essential feature of aesthetic experience is Jerome Stolnitz. Stolnitz traces this tradition to certain eighteenth-century British and Scottish thinkers, three of whom are noteworthy here: the Third Earl of Shaftesbury (Anthony Cooper), Francis Hutcheson, and Archibald Alison. Stolnitz suggests that the concept of disinterestedness relevant to aesthetic appreciation gradually developed over the eighteenth century in something like the following manner. First, Shaftesbury introduces the idea as a way of characterizing the experience of beauty, but initially his conception of disinterestedness is, relatively speaking, theoretically lightweight and primarily negative, not unlike the contemporary idea of a disinterested party – that is, an individual having no particular personal or selfish interest in a situation. Similarly, Shaftesbury contrasts a disinterested stance toward an object with using it for some purpose. Second, Hutcheson elaborates Shaftesbury's conception, expanding the idea so as to exclude not simply personal and self-serving utilitarian interest, but also interest of a more general nature and in particular cognitive interest. Third, the notion reaches its full theoretical development in the thoughts of Alison who treats disinterestedness as a particular "state of mind" which he characterizes in a famous passage as one of being "vacant and unemployed."[11] Contemporary theoreticians of disinterestedness, such as Stolnitz, have in many respects followed Alison's account.[12]

In light of even these extremely sketchy remarks about disinterestedness, it is clear that if aesthetic experience is analyzed in terms of disinterestedness in either the second or the third sense, then there is indeed a necessary conflict between aesthetic experience and any experience involving an understanding of meanings by means of knowledge gained through education or training. The second conception of disinterestedness attributed to Hutcheson appears to explicitly exclude any such cognitive dimensions from aesthetic experience, and Alison's vacant and unemployed state of mind would seemingly exclude that and much more. Moreover, in each case it is a conceptual matter: any experiences involving such understanding of meaning will not be aesthetic as a result of the very analysis of what it is to be an aesthetic experience. Thus, in light of these notions of disinterestedness, Twain's suggestion of a conflict between his two experiences could be explained and grounded after all. However, there are at least two problems with such an account. First, although Twain's idea of aesthetic experience is, as noted, a quite narrow, formalist conception that could be underwritten by one or the other of these theoretically vigorous notions of disinterestedness, there is no reason to think that Twain subscribes to such notions. Perhaps the uneducated passenger who sees nothing but pretty pictures has a vacant and unemployed mind, but, as also noted, Twain's aesthetic experience involves both more and less than pretty pictures. Second and more important, those considerations by means of which we previously demonstrated that the coherence of even the sparsest aesthetic appreciation requires a certain level and kind of knowledge facilitating understanding also count against the coherence of a conception of aesthetic experience constrained by these robust notions of disinterestedness. In short, perhaps a mind that is truly vacant and unemployed or even one that simply excludes all cognitive dimensions can have no aesthetic experience at all – or at least none such as Twain attributes to himself.

However, what about the essentially negative conception of disinterestedness attributed to Shaftesbury, the notion not so different from the contemporary, non-theoretical idea of disinterestedness? On the position that aesthetic experience is disinterested in this sense, an individual will apparently not have an aesthetic experience of an object if he or she has some particular personal or self-serving utilitarian interest in it. Even though this position is theoretically weaker than the other two, it is yet adequate for addressing Twain's conflict between his two experiences of the river. Here is where it is significant that Twain's second experience is a "steamboat pilot experience." Such an experience is a paradigm of the kind that this position claims is in conflict with aesthetic appreciation. The pilot's particular understanding of the meanings of the water, the understanding of hidden reefs, changing channels, dangerous shoals, new snags, and the like, is exactly the kind of self-serving utilitarian interest in the river that seemingly interferes with aesthetic experience, indeed perhaps even destroys the possibility of such experience. The pilot is only using the river for his or her own purposes, and this apparently makes aesthetic appreciation of it difficult, if not impossible. If nothing else, reading the language of the water in order to successfully utilize the river is for the pilot such an all-important and all-consuming interest that this will, at least in practice, if not in theory, exclude all other experiences of it. We must remember that the penalty for a pilot's failing to be single-mindedly obsessed by the language of the river can be the "killing of his steamboat."

As noted, however, this position is theoretically weaker than those involving the other two senses of disinterestedness. Part of the weakness lies in the fact that it need not be taken as an analysis of aesthetic experience requiring a commitment to a special state of mind. Rather, it may be taken as embodying only an empirical observation about the actual relationship between aesthetic appreciation and certain kinds of interests, such as those of a steamboat pilot. But given this, is it yet strong enough to really explain the conflict Twain finds between his two experiences of the river? Concerning this question, two points are relevant. On the one hand, as suggested above, the position, even if only empirical, yet clearly accounts for the conflict Twain reports, that is, his inability to aesthetically appreciate the river when he instead reads the language of the water as would a steamboat pilot. On the other hand, however, Twain claims that having mastered the language of the water, he has "lost something which could never be restored" to him in his lifetime. Is not, we might ask, an analysis of the nature of aesthetic experience rather than simply an empirical observation required to account for the finality of Twain's loss? In light of the strength of his claim, perhaps a conceptual point is required after all. However, if so, perhaps such a point about the role of steamboat pilot, rather than about aesthetic experience, is adequate. That is to say, to be a steamboat pilot, or at least to be a good one, may conceptually require that one single-mindedly engage in the all-consuming interest in the language of the water, which interest as a matter of fact excludes aesthetic experience of the river. If Twain has indeed lost something that can never be restored in his lifetime, perhaps it is only his ability, when confronted by the river, to take up any role other than the role of good steamboat pilot.

Conclusion

In conclusion, it must be admitted that our treatment of Twain's marvelous description of this one facet of his life on the Mississippi does not account for something we initially thought suggested by it, that is, a conflict between aesthetic experience and cognitive experience in general. However, although Twain's remarks suggest a general conflict, he is not, given his focus on the particular experience of the steamboat pilot, committed to any such conflict. Moreover, that he is not so committed is just as well, for we have discovered, first, in our consideration of the formalist theory of art, that defending any such general conflict leads to incoherence and, second, in our consideration of disinterestedness, that defending such a conflict is not necessary in order to explain observations such as Twain's. And if we can account for the observations of an individual as astute and perceptive as Twain without committing him to an incoherent and unnecessary position, that is probably some indication that we are on the right track.

Notes

1 Mark Twain, *Life on the Mississippi* [1883], New York, Penguin, 1984, pp. 94–6.
2 Clive Bell, *Art* [1913], New York, G.P. Putnam's Sons, 1958, p. 17. For a more general examination of formalism, formal qualities, and appreciation of nature, see "Formal Qualities in the Natural Environment," *Journal of Aesthetic Education*, 1979, vol. 13, pp. 99–114 (reproduced in this volume, Chapter 3). I also discuss these issues in "On the Possibility of Quantifying Scenic Beauty," *Landscape Planning*, 1977, vol. 4, pp. 131–72.
3 Bell, op. cit., p.17.
4 Ibid., p.17.
5 Ibid., p. 30.
6 Ibid., p. 45.
7 Ibid., p. 28.
8 Ibid., p. 27.
9 Ibid., p. 27.
10 Twain, op. cit., p. 94.
11 Archibald Alison, *Essays on the Nature and Principles of Taste* [1790], Third Edition, Edinburgh, Archibald Constable and Company, 1812, vol. I, p.10. The relevant sources for Shaftesbury and Hutcheson are, respectively, Anthony, Earl of Shaftesbury, *Characteristics of Men, Manners, Opinions, Times, etc.* [1711], John M. Robertson (ed.), Gloucester, Massachusetts, Peter Smith, 1963 and Francis Hutcheson, *An Inquiry Concerning Beauty, Order, Harmony, Design.* [1725], Peter Kivy (ed.), The Hague, Martinus Nijhoff, 1973.
12 See Jerome Stolnitz, "Of the Origins of 'Aesthetic Disinterestedness,'" *The Journal of Aesthetics and Art Criticism*, 1961, vol. 20, pp. 131–43. For Stolnitz's own position on disinterestedness, see Jerome Stolnitz, *Aesthetics and Philosophy of Art Criticism: A Critical Introduction*, Boston, Houghton Mifflin, 1960. For a fuller discussion of his position, see my "Appreciating Art and Appreciating Nature," in S. Kemal and I. Gaskell (eds) *Landscape, Natural Beauty, and the Arts*, Cambridge, Cambridge University Press, 1993, pp. 199–227 (reproduced in this volume, Chapter 7). I also consider the ramifications of disinterestedness in "Between Nature and Art" (in this volume, Chapter 8).

3

FORMAL QUALITIES IN THE NATURAL ENVIRONMENT

Formal qualities and formalism

Following up our introduction to formalism in Chapter 2, in this chapter I consider the significance of formal qualities – in particular, in the aesthetic appreciation and evaluation of the natural environment. The topic is of interest for at least two reasons. The first is that there is a marked emphasis on formal qualities in much of the current non-theoretical research in environmental aesthetics. The second is that recent philosophical discussion of the aesthetics of the natural environment suggests that the whole question of formal aesthetic qualities in the natural environment is more perplexing than it at first appears.

Initially, it is useful to introduce a notion of "formal qualities" somewhat more precise than the intuitive notion relied upon in Chapter 2. Writers in philosophical aesthetics have identified and classified the aesthetic qualities of objects in various ways. Two groupings of qualities about which there is some agreement, however, are sensory or sensuous qualities and formal or design qualities. The former are qualities of textures, colors, and lines of objects, such as the smoothness and luster of a polished stone or the sharpness and sparkle of a crystal. Sensory qualities are worth noting here in that their specification aids in the clarification of formal qualities. This is so in part because textures, lines, and colors combine in relations to create the shapes, patterns, and designs that constitute the perceived form of an object. It is the qualities of such forms, such as their being unified or chaotic, balanced or unbalanced, harmonious or confused, that I call formal qualities. It follows that formal qualities are qualities that objects or combinations of objects have in virtue of that which constitutes their forms. This includes not only their shapes, patterns, and designs, but also their textures, lines, and colors.

At this point it is also useful to further clarify the notion of formalism, especially as it is employed in the third section of this chapter. This can be done by reference to the above-mentioned aspects of objects. Concerning aesthetic appreciation, formalism holds that such appreciation is to be directed toward those aspects – textures, lines, colors and resultant shapes, patterns, and designs – that constitute the form of the object. Concerning aesthetic value, formalism holds that the formal qualities of an object, which it has in virtue of these aspects, are the only qualities relevant to the aesthetic value of that object. An object is aesthetically good in virtue of having formal qualities such as unity and balance – or more sophisticated variations such as "organic unity" or "variety in unity" – and aesthetically bad in virtue of having formal qualities such as disharmony or lack of integration.

Formal qualities in current work in environmental aesthetics

As suggested above, there is an emphasis on formal qualities in much current work in environmental aesthetics. The work in question originates in fields such as geography, forestry management, recreation planning, and landscape architecture. The overall aim of much of this work is to identify, classify, evaluate, or measure quantifiably the aesthetic dimension of the natural environment. For example, researchers hope or claim to identify and map large areas of "scenic value," to classify and appraise "visual landscape qualities," to evaluate and compare quantitatively the "aesthetic quality of different landscapes," and to "quantify scenic beauty."[1] To accomplish all this it is important to have a clear understanding of concepts such as "scenic value," "the aesthetic quality of the landscape," and "scenic beauty." It is in the understanding of these concepts that much research seemingly assumes an account of aesthetic appreciation and value which is heavily, if not exclusively, dependent upon formal qualities.

It is useful to briefly illustrate this point. The emphasis on formal qualities is attributable to researchers who describe and promote "the possibility and usefulness" of seeing "the managed forest in terms of design factors: form, contrast, distance, color, light, and angle of view."[2] This direction has apparently become the policy of the US National Forest Service. Its training document, which is the "basic text to illustrate the concepts, elements, and principles" of the Forest Service's landscape management program, describes "form, line, color, and texture" as the four main elements in any landscape and suggests principles such as "contrast," "axis," and "convergence" to analyze these elements. Concerning aesthetic appreciation, the aesthetic dimension of the environment – or, as it is put in the document, the "character of a landscape" – is defined as "the overall impression created by its unique combination of visual features (such as land, vegetation, water, and structures) as seen in terms of form, line, color, and texture." As for aesthetic value, it is suggested that the overall "scenic quality" of landscapes should be evaluated mainly in terms of the "variety" or "diversity" of these four formal elements.[3] Thus aesthetic appreciation is directed toward those elements that constitute the perceived form of a landscape, and aesthetic evaluation of "scenic quality" is elaborated by reference to certain formal qualities that arise from these elements.

The emphasis of formal qualities is also evident in research that attempts to measure the environment's aesthetic quality. An excellent example is the work of E. L. Shafer. He and his colleagues have attempted to measure quantitatively the "aesthetic quality of different landscapes" by trying to identify and measure "quantitative features in a landscape" which "affect its aesthetic appeal." Whether such an attempt can be successful is not important here; what is important is the methodology utilized. It involves dividing photographs of landscapes into different "zones" – sky and clouds, water, immediate trees and shrubs, immediate other features, distant trees and shrubs, and so forth – and measuring the perimeters and areas of such zones. The perimeter and area measurements are then fed into an equation in order to calculate the aesthetic preferability or aesthetic appeal of the landscapes.[4] What is significant about this methodology is its assumption that the "aesthetic quality of different landscapes" can be measured by means of measuring formal aspects of photographs. The zones to be measured are initially identified in terms of content (e.g., distant trees and shrubs), but the actual measurements are of certain shapes and lines and of the relationships holding between these shapes and lines. Being measurements of those elements that constitute form, they can at most indicate that a given combination of shapes and lines has formal qualities such as balance or proportion. In short, the attempt is to calculate the "aesthetic quality of different landscapes" in terms of form and formal qualities.

Background on the significance assigned to formal qualities

Some background is useful to help explain this emphasis on formal qualities. First to be mentioned is the rather recent popularity enjoyed by formalism in art and art criticism. As noted in Chapter 2, formalism as a theory of art and of art criticism was promoted early in this century by Clive Bell, Roger Fry, and other critics. Moreover, although the position has been subjected to extensive criticism by art critics and aestheticians, it has nonetheless had great influence on both the initial acceptance and the continued development of various degrees of abstraction in art. Hand in hand with these developments came the willingness and the ability on the part of both critics and the general public to respond to the more formal dimensions in art and to think of aesthetic appreciation and aesthetic value in terms of form and formal qualities. These developments were essentially healthy since previously the formal dimensions of art had perhaps not been adequately appreciated. And now vigorously growing new artistic traditions, different schools of criticism, and opposing theories of art have combined to achieve a balanced appreciation of the formal dimensions of art along with its other dimensions.

For our purposes, however, it is most important to note that formalism was promoted not only in art, but also concerning the natural environment. As we saw in Chapter 2, Bell, for example, suggests that when the artist sees non-art objects in "moments of inspiration" or "moments of aesthetic vision," he or she sees them as "pure forms." At one point he speculates:

> All of us, I imagine, do, from time to time, get a vision of material objects as pure forms. We see things as ends in themselves, that is to say, and at such moments it seems possible, and even probable, that we see them with the eye of an artist. Who has not, once at least in his life, had a sudden vision of landscape as pure form? For once, instead of seeing it as fields and cottages, he has felt it as lines and colors. In that moment has he not won from material beauty a thrill indistinguishable from that which art gives? And, if this be so, is it not clear that he has won from material beauty the thrill that, generally, art alone can give, because he has contrived to see it as a pure formal combination of lines and colors? May we go on to say that, having seen it as pure form, having freed it from all casual and adventitious interests, from all that it may have acquired from its commerce with human beings, from all its significance as a means, he has felt its significance as an end in itself?[5]

The implication here is clear: If we are to appreciate the natural environment aesthetically, we must "see it as a pure formal combination of lines and colors." Thus formalism was initially accepted as being as appropriate for the natural environment as it was for art. And although formalism concerning the natural environment developed somewhat more slowly than it did concerning art, the former (at least as a distinct emphasis on formal qualities) has stayed with us longer than the latter. I suspect this is so in part because, as noted in Chapter 1, until very recently there have not been many new traditions, different schools of criticism, or opposing theories with respect to the aesthetics of the natural environment. In this area formalism has not been placed in proper perspective, as it has in the arts; instead it has remained a relatively uncontested and popular point of view. As a result, formalism is not only taken for granted as the basis for much empirical research, but a version of it also emerges in some of the more theoretical discussions. Thus we find individuals today saying essentially what Bell says in the above quote. For example, in discussing a criterion for scenic beauty one such individual, writing exactly a half century after Bell, states:

> Preoccupation with purpose is in fact no aid, but a deterrent to landscape appreciation. The man who has to consider what things are used for is least likely to note their shapes, colors and patterns. To exalt his judgment is to promote a complacent inattention to appearance, an abnegation of esthetic response.[6]

A second factor that may explain the assumed importance of formal qualities in the aesthetics of the natural environment is also closely related to a development in art and art appreciation. This is the rise of landscape painting and its effects on aesthetic appreciation. This development is at least in part responsible for a certain mode of perceiving and aesthetically appreciating the natural environment, the core of which is the landscape model of nature appreciation introduced in Chapter 1. This mode of appreciation is clearly evident in what is sometimes called the "scenery cult."[7] Essentially, it encourages perceiving and appreciating the natural environment as if it were a landscape painting, that is, as a certain kind of prospect seen from a specific standpoint. The mode of appreciation was exemplified in the eighteenth and nineteenth centuries in the use of the "Claude-glass," a small, tinted, convex mirror named for the landscape artist. Tourists used the mirror to view the landscape, finding that it, as a travel book of the day put it, removes objects "great and near" "to a due distance, and shews them in the soft colors of nature, and most regular perspective the eye can perceive;" the glass gives "the finished picture, in highest coloring, and just perspective."[8] Likewise, modern tourists reveal their preferences for this same mode of appreciation by frequenting "scenic viewpoints," where the actual space between tourists and the prescribed "view," as well as the position of the viewpoint itself, often enhances elements of color and perspective. Moreover, modern tourists also seek "the finished picture, in highest coloring, and just perspective." It can be found in the "scene" framed in the viewfinders of their cameras and in the resultant Kodachrome slides, as well as in the ever-popular formally composed postcards of the "scene." The point is that, as one commentator puts it, "the taste has been for a view, for scenery, [for] a prospect."[9]

The significance of the scenery cult for our interests here is that the related mode of appreciation accentuates overall form and formal qualities and thus implicitly guides the appreciator towards these qualities in the appreciation of the natural environment. In the carefully composed landscape paintings – and even in their poor cousins, the postcard reproductions – form-constituting elements, resultant overall form, and formal qualities such as balance, proportion, and organization are of considerable importance. These formal aspects are rightly appreciated in, and contribute significantly to the aesthetic value of, such art works and reproductions. Consequently, when tourists attempt to perceive and appreciate the natural environment as if it were a landscape painting, they are guided toward the formal dimensions. The Claude-glass, the camera viewfinder, and the scenic viewpoint all help make the environment look more like landscape art as they tone down harsh contrasts, make detail less evident, accentuate overall relations and patterns, and more or less provide limits for the scene. But in all this the form and formal qualities of the environment are enhanced at the expense of the natural environment's other aesthetically significant qualities. In this way the scenery cult not only promotes the aesthetic appreciation of the natural environment (as has been widely recognized); it also promotes a certain kind of aesthetic appreciation – appreciation with a formalist bias.

The way in which the scenery cult promotes formal qualities in the aesthetic appreciation of the natural environment has an interesting parallel in current research. The difference is that this research emphasizes formal qualities not by adopting a specific mode of appreciation, but by utilizing a certain kind of substitute for the natural environment – the photograph. Much research, such as that of Shafer mentioned above, uses photographs of the environment as "stand-ins" that make research easier, quicker, and less expensive. It is important to note, however, that the use of photographs affects aesthetic appreciation. The photograph, like the Claude-glass and the camera viewfinder, makes detail less evident, accentuates overall relations and patterns, and provides limits for the scene. Moreover, the components of a photograph are shapes, colors, lines, and patterns and not the trees, shrubs, and rocks that compose the natural environment. Thus it is easier and more "natural" for individuals to perceive the natural environment in terms of shapes, lines, colors and patterns when they are presented with a photograph of it than when they are in the environment itself. The upshot is that if one plans and executes research in environmental aesthetics in terms of photographs rather than the actual environment, a concentration on form-constituting elements and formal qualities becomes a natural approach. Moreover, such an approach appears even more natural given the historical prominence of artistic formalism and the larger context of the scenery cult with its attendant mode of appreciation.

Formal qualities in the natural environment

The preceding section suggests that the emphasis on formal qualities in the natural environment is related to and influenced by traditional formalism in art. In view of this, one way of approaching the issue of the significance of formal qualities in the aesthetic appreciation and evaluation of the natural environment might be to consider the plausibility of formalism concerning art. I make some suggestions along these lines in Chapter 2, but do not take this tack here, for I believe enough has been said in that chapter and in the literature in general to show the implausibility of such a formalism.[10] Rather, I approach the issue of the significance of formal qualities by considering what seems to be implicitly assumed by the views and positions discussed above. This is the assumption that formal qualities have at least the same place and importance in our aesthetic appreciation and evaluation of the natural environment as they have in the appreciation and evaluation of art. Consideration of this, in turn, requires examining some differences between the natural environment and works of art.

Further consideration of the scenery cult and the use of photographs is helpful in this regard as these two factors tend to obscure some differences between the natural environment and works of art. Let us first consider the scenery cult. This cult, as noted, puts the accent on perceiving and appreciating the natural environment as if it were a landscape painting, that is, as a certain kind of prospect seen from a specific standpoint. The mode of appreciation involved divides the environment into scenes or blocks of scenery, each of which is to be viewed from a particular point by a viewer who is separated by an appropriate spatial (and emotional) distance. Thus a drive through a national park is not unlike a stroll through a gallery of landscape paintings; we stop at each viewpoint as we pause before each painting.

Seen in this light, the scenery cult seems to have some rather questionable aspects. These begin to become clear when it is realized that viewing the environment as if it were a landscape painting means viewing it as a static, essentially two-dimensional representation. This requires the reduction of the environment to a scene or view. But the environment is not a scene, not a representation, not static, and not two-dimensional. The point is that the mode of appreciation of the scenery cult requires the appreciation of the environment not as what it is and with the qualities it has, but as something which it is not and with qualities it does not have. Consequently, when focused on the natural environment, the scenery cult provides a mode of appreciation that is inappropriate to the actual nature of the object of appreciation. This lack of "fit" between the mode and the object of appreciation is certainly an aesthetic deficiency of the scenery cult.[11] A similar case of deficiency would arise if we appreciated people aesthetically as if they were only photographs or portraits of themselves. More common cases are viewing sculpture as if it were painting or observing a "happening" as if it were a play. The appropriate mode of appreciation for sculpture calls on us to touch it, to walk around it, to physically position ourselves in relation to it; for the happening, to participate in it actively, to be part of it.

Because of this deficiency, the scenery cult obscures a significant difference between landscape paintings and the natural environment. A painting, unlike the natural environment, is to be appreciated from a specific point external to it and with only one sense, the sense of sight. This difference is also obscured by the other factor mentioned previously, that is, the use of photographs in empirical research. When a photograph is used as a "stand-in" for the actual environment, the mode of appreciation of the scenery cult is in essence forced on the appreciator. He or she views the object of appreciation only with his or her sense of sight and from a specific external point, because the "stand-in," the photograph, must be so viewed. There is no point in smelling or touching the photograph, or in walking around it; and one cannot, of course, walk around in it. But with the actual object of appreciation – the natural environment – all these things are not only possible, they are also appropriate and have a very real point. The point is that they contribute to the full, rich aesthetic appreciation that the natural environment both allows and encourages. Whether we realize it or not, we are necessarily in the natural environment when we appreciate it aesthetically, and often we are moving around in it and around parts of it. Ronald Hepburn puts the point as follows:

> On occasion he [the spectator] may confront natural objects as a static, disengaged observer; but far more typically the objects envelop him on all sides. In a forest, trees surround him; he is ringed by hills, or he

> stands in the midst of a plain. If there is movement in the scene, the spectator may himself be in motion and his motion may be an important element in his aesthetic experience.[12]

Thus aesthetic appreciation of the natural environment is not simply a matter of looking at objects or "views" from a specific point. Rather, it is being "in the midst" of them, moving in regard to them, looking at them from any and every point and distance and, of course, not only looking, but also smelling, hearing, touching, feeling. It is being in the environment, being a part of the environment, and reacting to it as a part of it. It is such active, involved aesthetic appreciation, rather than the formal mode of appreciation nurtured by the scenery cult and encouraged by photographs, that is appropriate to the natural environment.[13]

Before the ramifications of this point are pursued, another difference between art and the natural environment – one having to do with the objects of appreciation rather than with appropriate modes of appreciation – needs to be examined. Again the scenery cult is revealing. It encourages perceiving and appreciating the natural environment as if it were a landscape painting, and it traditionally depended on devices such as the Claude-glass and more recently on the camera viewfinder to achieve this. Such devices are important because they help "frame," and compose within the frame, a section of the actual environment. In a similar way, landscape painters rely on the diminishing glass and the motive cutter in deciding on landscape subjects. The latter device is said to "help to determine the exact proportions, or limits of a subject best" and is equipped with "a sliding crossbar to be moved until the right proportions have been found."[14] The point is that if we wish to discern a potential landscape painting in the natural environment or to perceive and appreciate that environment as if it were a landscape painting, the relevant part of the environment must be framed or cut apart from the rest in some manner. This is because of an obvious but important difference between traditional art objects and the natural environment: the former, but not the latter, are framed or delineated in some formal way.[15] This difference is not only de-emphasized by the scenery cult, but is also obscured by the use of photographs and slides as both a means of conducting empirical research and a way of (vicariously) appreciating the natural environment. Photographs and slides, like traditional works of art, are framed and, depending on the skill of the photographer, are poorly or well composed within that frame.

The presence or absence of a frame has significance for formal qualities. For example, traditional art objects are (or are not) unified or balanced within their frames and in relation to their frames. This is obvious not only from the art objects themselves, but also from the way individuals are taught to create them. Art students are urged to think of their works not just as designs or compositions *simpliciter*, but as being designed into a frame or composed within a frame. They are taught to pay attention to the "negative spaces" within their works, that is, the shapes formed by the outside edges of their designs and shapes and the inside edges of the frame. The point is that when an object such as a traditional work of art is framed in some way, formal qualities are in large part determined by the frame. Consequently, since the frame is static and the object is appreciated within its frame, its formal qualities are an important determinate aspect of the work itself. Formal qualities in art, therefore, can be easily appreciated and evaluated; in fact they must be in order to achieve full appreciation and correct evaluation of the work.

As for the natural environment, consider the following example. Near Jasper, Alberta, Canada, there are six cabins with large windows overlooking a majestic, mountain-ringed lake. To one standing outside these cabins, the natural environment appears neither balanced nor unbalanced, neither unified nor chaotic. But to one sitting inside one of them across from a window, the view framed by the window is striking in its balance and unity; to one standing across from the window, however, the view is partly destroyed by a lack of balance resulting from the top of the window frame obscuring the peak of the highest mountain. This example suggests a distinction between the natural environment and any particular framed view of that environment. It further suggests that it is the latter rather than the former that can be properly said to have formal qualities. As the example shows, the natural environment itself only appears to have formal qualities when, in one way or another, a person imposes a frame upon it and thus formally composes the resultant view. And in such a case it is the framed view that has the qualities. Moreover, the formal qualities of such a view will vary depending upon the frame and the viewer's position. In fact, for any part of the environment there are a near infinity of possible frames and positions that would in turn produce a near infinity of different formal qualities in a near infinity of different views. The natural environment itself might at best be seen as only a "source" of different views, each with somewhat different formal qualities. Given this situation, it might be concluded that insofar as the natural environment has formal qualities, they are qualities with a certain built-in indeterminateness, making them difficult to appreciate and accordingly a relatively insignificant aspect of the aesthetic appreciation and evaluation of that environment.[16]

I think that the above position is essentially correct as far as it goes. However, I also think that a somewhat stronger position – that the natural environment as such does not possess formal qualities – is worth considering. The argument for this stronger position can be developed from the earlier conclusion that the appropriate mode of aesthetic appreciation of the natural environment is the active, involved appreciation of one who is in the environment, being a part of and reacting to it. This conclusion is significant in that it is inconsistent with framing the environment. In framing a section of the environment, one must become a static observer who is separate from that section and who views it from a specific external point. But one cannot be engaged in the appropriate active, involved appreciation while maintaining the static, external point of view required for framing. In short, one cannot both be in the environment that one appreciates and frame that environment; if one appreciates the environment by being in it, it is not a framed environment that one appreciates. Consequently, framing itself must be seen as an inappropriate way of attempting to aesthetically appreciate and evaluate the natural environment. Without imposing a frame on the natural environment, however, it is not possible to see it as having formal qualities. Framing may allow the environment to be seen as having certain indeterminate formal qualities; without framing, the environment seemingly has no formal qualities, indeterminate or otherwise. Thus, when appreciated in the appropriate mode, the natural environment as such has no formal qualities to appreciate.

I do not think that the above is a very surprising or implausible conclusion. Seeing its plausibility only requires reflection on the nature of formal qualities and on the nature of the natural environment. The former predominate in those areas of art where imposed restraints and limits allow them to be made manifest. The natural environment, by contrast, has a certain openness and indeterminateness that makes it an unlikely place to find formal qualities. This has been recognized by landscape painters who have emphasized that in order to create a landscape painting (to which formal qualities are, of course, significant), an artist cannot simply copy the landscape faithfully. Rather, he or she must compose the work such that formal qualities arise. The landscape painter Adrian Stokes, in a text on landscape painting, offers the following advice:

> A subject in nature, which has been chosen on account of the appeal it makes to the artist, can rarely be cut out from its surroundings in such a way that the composition is not in need of change or modification. It must be invested with some quality of rhythm or symmetry. To copy each part faithfully, even in truthful relation to other parts, is not sufficient. All things must be considered together as forming portions of one design, one harmonious whole.[17]

Santayana, a philosopher with a sense of beauty, also notes that the landscape itself "has no real unity." He makes a point similar to that of Stokes, although his philosophical bent draws him to a further conclusion:

> The natural landscape is an indeterminate object: it almost always contains enough diversity to allow the eye a great liberty in selecting, emphasizing, and grouping its elements, and it is furthermore rich in suggestion and in vague emotional stimulus. A landscape to be seen has to be composed...then we feel that the landscape is beautiful. This is a beauty dependent on reverie, fancy, and objectified emotion. The promiscuous natural landscape cannot be enjoyed in any other way.[18]

The implication of Santayana's remarks is clear (and consistent with what I have argued). The natural environment cannot be appreciated and valued in terms of formal beauty, that is, the beauty of formal qualities; rather, it must be appreciated and valued in terms of its other aesthetic dimensions – its various non-formal aesthetic qualities and, according to Santayana, especially its expressive emotional qualities: "The promiscuous natural landscape cannot be enjoyed in any other way."[19]

Conclusion

If Santayana's insight and what I have argued are correct, the conclusions to be drawn are clear enough. The assumption that formal qualities have the same place and importance in the aesthetic appreciation and evaluation of the natural environment as they do in the appreciation and evaluation of art must be abandoned, for the natural environment is such that formal qualities have relatively little place and importance. Consequently, the emphasis should be removed from formal qualities and, as Santayana suggests, placed on the non-formal aesthetic qualities of the natural environment. It is thus evident that much current research, such as that noted in the second section of this chapter, which emphasizes formal qualities to the near exclusion of other aesthetic qualities, is not only progressing on an inappropriate basis, but is misleading concerning the nature of aesthetic appreciation and evaluation of the natural environment. If we are fully to appreciate our natural environment and evaluate its aesthetic dimension wisely, then we, and those carrying out research, must look in other directions.

Notes

1 Aims such as these are evident in much research in this area. See, for example, Elwood L. Shafer, Jr. and James Meitz, *It Seems Possible to Quantify Scenic Beauty in Photographs*, USDA Forest Service Research Paper NE-162, Upper Darby, Pennsylvania, USDA Northeastern Forest Experiment Station, 1970; Luna B. Leopold, "Landscape Esthetics: How to Quantify the Scenics of a River Valley," *Natural History*, 1969, vol. 78, pp. 36–45; G. Wright, "Appraisal of Visual Landscape Qualities in a Region Selected for Accelerated Growth," *Landscape Planning*, 1974, vol. 1, pp. 307–27; and K. D. Fines, "Landscape Evaluation: A Research Project in East Sussex," *Regional Studies*, 1968, vol. 2, pp. 41–55. For a general review of the literature in this area together with a good selected bibliography, see E. H. Zube, "Scenery as a Natural Resource: Implications of Public Policy and Problems of Definition, Description, and Evaluation," *Landscape Architecture*, 1973, vol. 63, pp. 125–32. I develop a fuller critique of this research, focusing especially on the work of Shafer and his colleagues, in "On the Possibility of Quantifying Scenic Beauty," *Landscape Planning*, 1977, vol. 4, pp. 131–72.

2 Robert H. Twiss and R. Burton Litton, "Research on Forest Environmental Design," *Proceedings: Society of American Foresters*, Washington, DC, Society of American Foresters, 1967, p. 209. In the same proceedings, see also Litton and Twiss, "The Forest Landscape: Some Elements of Visual Analysis," pp. 212–14.

3 USDA Forest Service, *National Forest Landscape Management, Volume I*, USDA Handbook no. 434, Washington, D.C., Government Printing Office, 1972, pp. 7, 23–47; and *National Forest Landscape Management, Volume II*, USDA Handbook no. 462, Washington, DC, Government Printing Office, 1973, pp.2, 12.
4 For a full account of this methodology, see Shafer and Meitz, op. cit., and Elwood L. Shafer, Jr., John F. Hamilton, Jr., and Elizabeth A. Schmidt, "Natural Landscape Preferences: A Predictive Model," *Journal of Leisure Research*, 1969, vol. 1, pp 1–19.
5 Clive Bell, *Art* [1913], New York, G. P. Putnam's Sons, 1958, p. 45.
6 David Lowenthal, "Not Every Prospect Pleases: What Is Our Criterion for Scenic Beauty?," *Landscape*, 1962–63, vol. 12, pp. 19–23. This quote is from p. 22.
7 See Christopher Hussey, *The Picturesque*, London, G. P. Putnam's Sons, 1927 or E. W. Manwaring, *Italian Landscape in XVIII Century England*, New York, Oxford University Press, 1925. For a brief discussion, see R. Rees, "The Scenery Cult: Changing Landscape Tastes over Three Centuries," *Landscape*, 1975, vol. 19, pp. 39–47.
8 Thomas West, *Guide to the Lakes*, London, 1778, as quoted in J. T. Odgen, "From Spatial to Aesthetic Distance in the Eighteenth Century," *Journal of the History of Ideas*, 1974, vol. 35, pp. 63–78. This quote is from pp.66–7.
9 R. Rees, "The Taste for Mountain Scenery," *History Today*, 1975, vol. 25, pp. 305–12. This quote is from p. 312. I consider the mode of appreciation outlined in this paragraph in more detail in "Appreciation and the Natural Environment," *The Journal of Aesthetics and Art Criticism*, 1979, vol. 37, pp. 267–76 (reproduced in this volume, Chapter 4).
10 Concerning art, the case against formalism is nicely developed in John Hospers, *Meaning and Truth in the Arts*, Chapel Hill, University of North Carolina Press, 1946, chap. 4 and Walter Abell, *Representation and Form: A Study of Aesthetic Values in Representational Art*, New York, Charles Scribner's, 1936, *passim*. Concerning the natural environment, a significant consideration seems to be the role played in our appreciation and evaluation by non-formal aesthetic qualities such as expressive qualities and qualities such as gracefulness, delicacy, and garishness. On this point, see, for example, R. W. Hepburn, "Aesthetic Appreciation of Nature," in H. Osborne (ed.) *Aesthetics in the Modern World*, London, Thames and Hudson, 1968; R. A. Smith and C. M. Smith, "Aesthetics and Environmental Education," *Journal of Aesthetic Education*, 1970, vol. 4, pp. 125–40; M. Sagoff, "On Preserving the Natural Environment," *Yale Law Journal*, 1974, vol. 84, pp. 205–67; Fines, op. cit.; Yi-Fu Tuan, "Man and Nature; An Eclectic Reading" *Landscape*, 1966, vol. 15, pp. 30–6.
11 Cf. Hepburn, op. cit., pp. 60–5.
12 Ibid., p. 51.
13 I develop the ramifications of these observations more fully in "Appreciation and the Natural Environment," op. cit. (reproduced in this volume, Chapter 4).
14 Adrian Stokes, *Landscape Painting*, London, Seeley, Service & Co., 1925, p. 214.
15 Cf. Hepburn, op. cit., pp. 51–2.
16 This appears to be the position of Hepburn, ibid., pp. 52–60 and especially p. 59, and of Smith and Smith, op. cit., pp. 134–7. Hepburn notes that "natural aesthetic objects" lack "full determinateness and stability." This leads him to the conclusion that "any aesthetic quality in nature is always provisional," but he further argues that certain formal qualities highly valued in the arts – for example, unity – are simply not equally significant in the aesthetic appreciation of nature and, consequently, that the aesthetic appreciation of nature cannot be restricted to consideration of formal qualities (pp. 52–3). Smith and Smith conclude that the "feature of being an ingredient in the perceptual field...makes it difficult to experience in nature the kind of formal unity, balance, proportion or self-enclosedness that are so important to the proper perception

of works of art," which explains "why it would be inappropriate...to respond to nature as one would to a work of art, that is, in terms of its formal properties of design" (p. 134).

17 Stokes, op. cit., p. 64.

18 George Santayana, *The Sense of Beauty* [1896], New York, Collier, 1961, p. 99.

19 Cf. Sagoff, op. cit., *passim.* Although he does not discuss formal qualities as such, Sagoff seemingly holds a position similar to that of Santayana and to the one which we have reached here. He attempts "to explain the aesthetic value of natural environments" by considering "only the expressive qualities of these environments and not their beauty, considered formally" (p. 248).

4

APPRECIATION AND THE NATURAL ENVIRONMENT

The appreciation of art

With art objects there is a straightforward sense in which we know both what and how to aesthetically appreciate. We know *what* to appreciate in that, first, we can distinguish a work and its parts from that which is not it nor a part of it. And, second, we can distinguish its aesthetically relevant aspects from its aspects without such relevance. We know that we are to appreciate the sound of the piano in the concert hall and not the coughing that interrupts it; we know that we are to appreciate that a painting is graceful, but not that it happens to hang in the Louvre. In a similar vein, we know *how* to appreciate in that we know what "acts of aspection" to perform concerning different works. Paul Ziff says:

> ...to contemplate a painting is to perform one act of aspection; to scan it is to perform another; to study, observe, survey, inspect, examine, scrutinize, etc., are still other acts of aspection...I survey a Tintoretto, while I scan an H. Bosch. Thus I step back to look at the Tintoretto, up to look at the Bosch. Different actions are involved. Do you drink brandy in the way you drink beer?[1]

It is clear that we have such knowledge of what and how to aesthetically appreciate. It is, I believe, also clear what the grounds are for this knowledge. Works of art are our own creations; it is for this reason that we know what is and what is not a part of a work, which of its aspects are of aesthetic significance, and how to appreciate them. We have made them for the purpose of aesthetic appreciation; in order for them to fulfil this purpose this knowledge must be accessible. In making an object we know what we make and thus its parts and its purpose. Hence in knowing what we make we know what to do with that which we make. In the more general cases the point is clear enough: in creating a painting, we know that what we make is a painting. In knowing this we know that it ends at its frame, that its colors are aesthetically important, but where it hangs is not, and that we are to look at it rather than, say, listen to it.

All this is involved in what it is to be a painting. Moreover, this point holds for more particular cases as well. Works of different particular types have different kinds of boundaries, have different foci of aesthetic significance, and perhaps most important demand different acts of aspection. In knowing the type we know what and how to appreciate. Ziff again:

> Generally speaking, a different act of aspection is performed in connection with works belonging to different schools of art, which is why the classification of style is of the essence. Venetian paintings lend themselves to an act of aspection involving attention to balanced masses: contours are of no importance, for they are scarcely to be found. The Florentine school demands attention to contours, the linear style predominates. Look for light in a Claude, for color in a Bonnard, for contoured volume in a Signorelli.[2]

I take the above to be essentially beyond serious dispute, except as to the details of the complete account. If it were not the case, our complementary institutions of art and of the aesthetic appreciation of art would not be as they are. We would not have the artworld that we do. But the subject of this chapter is not art nor the artworld. Rather it is the aesthetic appreciation of nature. The question I wish to investigate is the question of what and how to aesthetically appreciate concerning the natural environment. It is of interest since the account that is implicit in the above remarks, and that I believe to be the correct account for art, cannot be applied to the natural environment without at least some modification. Thus initially the questions of what and how to appreciate concerning nature appear to be open questions.

Some artistic models for the appreciation of nature

In this section I consider the two artistic paradigms that were introduced in Chapter 1: the object model and the landscape model. These paradigms seem *prima facie* applicable as models for the appreciation of the natural environment. In considering them I follow tradition in that these paradigms are ones that have been offered as or assumed to be appropriate models for the appreciation of nature. However, as suggested in Chapter 1, these models are not as promising as they initially appear to be.

The first artistic paradigm is the object model. In the artworld non-representational sculpture best fits this model of appreciation. When we appreciate such sculpture we appreciate it as the actual physical object that it is. The qualities to be aesthetically appreciated are the sensuous and design qualities of the actual object and perhaps certain abstract expressive qualities. The sculpture need not represent anything external to itself; it need not lead the appreciator beyond itself; it may be a self-contained aesthetic unit. Consider a Brancusi sculpture – for example, the famous *Bird In Space* (1919). It has no representational connections with the rest of reality and no relational connections with its immediate surroundings and yet it has significant aesthetic qualities. It glistens, has balance and grace, and expresses flight itself.

Clearly it is possible to aesthetically appreciate an object of nature in the way indicated by this model. For example, we may appreciate a rock or a piece of driftwood in the same way as we appreciate a Brancusi sculpture: we actually or contemplatively remove the object from its surroundings and dwell on its sensuous and design qualities and its possible expressive qualities. Moreover, there are considerations that support the plausibility of this model for appreciation of the natural environment. First, natural objects are in fact often appreciated in precisely this way: mantel pieces are littered with pieces of rock and driftwood. Second, the model fits well with one feature of natural objects: such objects, like the Brancusi sculpture, do not have representational ties to the rest of reality. Third and most important, the model involves an accepted, traditional aesthetic approach. As Francis Sparshott notes: "When one talks of the aesthetic this or that, one is usually thinking of it as entering into a subject/object relation."[3]

In spite of these considerations, however, I think there are aspects of the object model that make it inappropriate for nature. Santayana, in discussing the aesthetic appreciation of nature (which he calls the love of nature) notes that certain problems arise because the natural landscape has "indeterminate form." He then observes that although the landscape contains many objects which have determinate forms, "if the attention is directed specifically to them, we have no longer what, by a curious limitation of the word, is called the love of nature."[4] I think this limitation is not as curious as Santayana seems to think it is. The limitation marks the distinction between appreciating nature and appreciating the objects of nature. The importance of this distinction is seen by realizing the difficulty of appreciating nature by means of the object model. For example, on one understanding of the object model, the objects of nature when so appreciated become "ready mades" or "found art." The artworld grants "artistic enfranchisement" to a piece of driftwood just as it has to Duchamp's urinal, *Fountain* (1917), or to the real Brillo cartons discussed by Arthur Danto.[5] If this magic is successful the result is art. Questions of what and how to aesthetically appreciate are answered, of course, but concerning art rather than nature; the appreciation of nature is lost in the shuffle. Appreciating sculpture which was once driftwood is no closer to appreciating nature than is appreciating a totem pole that was once a tree or a purse that was once a sow's ear. In all such cases the conversion from nature to art (or artifact) is complete; only the means of conversion are different.

There is, however, another understanding of how the object model applies to the objects of nature. On this understanding natural objects are simply (actually or contemplatively) removed from their surroundings, but they do not become art, they remain natural objects. Here we do not appreciate the objects qua art objects, but rather qua natural objects. We do not consider the rock on our mantel a readymade sculpture, we consider it only an aesthetically pleasing rock. In such a case, as the example of non-representational sculpture suggests, our appreciation is limited to the sensuous and design qualities of the natural object and perhaps a few abstract expressive qualities: our rock has a wonderfully smooth and gracefully curved surface and expresses solidity.

The above suggests that, even when it does not require natural objects to be seen as art objects, the object model imposes a certain limitation on our appreciation of natural objects. The limitation is the result of the removal of the object from its surroundings that the object model requires in order even to begin to provide answers to questions of what and how to appreciate. But in requiring such a removal the object model becomes problematic. The object model is most appropriate for those art objects which are self-contained aesthetic units. These objects are such that neither the environment of their creation nor the environment of their display are aesthetically relevant: the removal of a self-contained art object from its environment of creation will not vary its aesthetic qualities and the environment of display of such an object should not affect its aesthetic qualities. However, natural objects possess what we might call an organic unity with their environments of creation: such objects are a part of and have developed out of the elements of their environments by means of the forces at work within those environments. Thus the environments of creation are aesthetically relevant to natural objects. And for this reason the environments of display are equally relevant in virtue of the fact that these environments will be either the same as or different from the environments of creation. In either case the aesthetic qualities of natural objects will be affected. Consider again our rock: on the mantel it may seem wonderfully smooth and gracefully curved and expressive of solidity, but in its environment of creation it will have more and different aesthetic qualities – qualities that are the product of the relationship between it and its environment. It is here expressive of the particular forces that shaped and continue to shape it and displays for aesthetic appreciation its place in and its relation to its environment. Moreover, depending upon its place in that environment it may not express many of those qualities, for example, solidity – that it appears to express when on the mantle.

I conclude that the object model, even without changing nature into art, faces a problem as a paradigm for the aesthetic appreciation of nature. The problem is a dilemma: either we remove the object from its environment or we leave it where it is. If the object is removed, the model applies to the object and suggests answers to the questions of what and how to appreciate. But the result is the appreciation of a comparatively limited set of aesthetic qualities. On the other hand, if the object is not removed, the model seemingly does not constitute an adequate model for a very large part of the appreciation that is possible. Thus it makes little headway with the what and how questions. In either case the object model does not provide a successful paradigm for the aesthetic appreciation of nature. It appears after all not a very "curious limitation" that when our attention is directed specifically toward the objects in the environment it is not called the love of nature.

The second artistic paradigm for the aesthetic appreciation of nature is that which I call in Chapter 1 the landscape model. In the artworld this model of appreciation is illustrated by landscape painting; in fact the model probably owes its existence to this art form. In one of its favored senses "landscape" indicates a prospect – usually a grand prospect – seen from a specific standpoint and distance; a landscape painting is frequently a representation of such a prospect.[6] When aesthetically appreciating landscape paintings (or any representative paintings, for that matter) the emphasis is not on the actual object (the painting) nor on the object represented (the actual prospect); rather it is on the representation of the object and its represented features. Thus in landscape painting the appreciative emphasis is on those qualities which play an essential role in representing a prospect: visual qualities related to coloration and overall design. These are the qualities that are traditionally significant in landscape painting and that are the focus of the landscape model of appreciation. We thus have a model of appreciation that encourages perceiving and appreciating nature as if it were a landscape painting, as a grand prospect seen from a specific standpoint and distance. It is a model that centers attention on those aesthetic qualities of color and design that are seen and best seen at a distance.

As noted in Chapter 3 in connection with formalism, the landscape model as embodied in the scenery cult has been historically significant in our aesthetic appreciation of nature.[7] The landscape model is the direct descendant of the scenery cult's key concept: the picturesque. This term literally means 'picture-like' and indicates a mode of appreciation by which the natural world is divided into scenes, each aiming at an ideal dictated by art, especially landscape painting. The concept guided the aesthetic appreciation of eighteenth-century tourists as they pursued picturesque scenery with the help of the "Claude-glass." Named after landscape artist Claude Lorrain, this small, tinted, convex mirror helped tourists see the landscape as they would art. Thomas West's popular guidebook to the Lake District (first published in 1778) says of the glass:

> ...where the objects are great and near, it removes them to a due distance, and shews them in the soft colors of nature, and most regular perspective the eye can perceive, art teach, or science demonstrate...to the glass is reserved the finished picture, in highest coloring, and just perspectives.[8]

And as also noted in Chapter 3, the influence of the scenery cult continues into the present as modern tourists similarly reveal their preferences for the landscape model of appreciation by frequenting "scenic viewpoints" where the actual space between tourists and the prescribed "view" often constitutes "a due distance" which aids the impression of "soft colors of nature, and the most regular perspective the eye can perceive, art teach, or science demonstrate." And the "regularity" of the perspective is often enhanced by the positioning of the viewpoint itself. Moreover, modern tourists also desire "the finished picture, in highest coloring, and just perspective"; whether this be the "scene" framed and balanced in the camera's viewfinder, the result of this in the form of a Kodacolor print, or the "artistically" composed postcard and calendar reproductions of the "scene" that often attract more appreciation than that which they "reproduce." Geographer R. Rees has described the situation as follows:

> ...the taste has been for a view, for scenery, not for landscape in the original Dutch – and present geographical – meaning of term, which denotes our ordinary, everyday surroundings. The average modern sightseer, unlike many of the Romantic poets and painters who were accomplished naturalists, is interested not in natural forms and processes, but in a prospect.[9]

It is clear that in addition to being historically important, the landscape model, like the object model, gives us at least initial guidelines as to what and how to appreciate in nature. We are to appreciate the natural environment as if it were a landscape painting. The model requires dividing the environment into scenes or blocks of scenery, each of which is to be viewed from a particular point by a viewer separated by the appropriate spatial (and emotional?) distance. A drive through the country is not unlike a walk through a gallery of landscape paintings. When seen in this light, this model of appreciation causes a certain uneasiness in a number of thinkers. Some, such as ecologist Paul Shepard, seemingly believe this kind of appreciation of the natural environment so misguided that they entertain doubts about the wisdom of any aesthetic approach to nature.[10] Others find the model to be ethically suspect. For example, after pointing out that the modern sightseer is interested only in a prospect, Rees concludes:

> In this respect the Romantic Movement was a mixed blessing. In certain phases of its development it stimulated the movement for the protection of nature, but in its picturesque phase it simply confirmed our anthropocentrism by suggesting that nature exists to please as well as to serve us. Our ethics, if the word can be used to describe our attitudes and behavior toward the environment, have lagged behind our aesthetics. It is an unfortunate lapse which allows us to abuse our local environments and venerate the Alps and the Rockies.[11]

What has not been as generally noted, however, is the point emphasized in the previous chapter. This is that this model of appreciation is suspect not only on ethical grounds, but also on aesthetic grounds. The model requires us to view the environment as if it were a static representation that is essentially "two dimensional." It requires the reduction of the environment to a scene or view. But what must be kept in mind is that the environment is not a scene, not a representation, not static, and not two dimensional. The point is that the model requires the appreciation of the environment not as what it is and with the qualities it has, but rather as something that it is not and with qualities it does not have. The model is in fact inappropriate to the actual nature of the object of appreciation. Consequently it not only, as with the object model, unduly limits our appreciation – in this case to visual qualities related to color and overall design – it also misleads it. Ronald Hepburn puts this point in a general way:

> Supposing that a person's aesthetic education...instills in him the attitudes, the tactics of approach, the expectations proper to the appreciation of art works only, such a person will either pay very little aesthetic heed to natural objects or else heed them in the wrong way. He will look – and of course look in vain – for what can be found and enjoyed only in art.[12]

An environmental model for the appreciation of nature

I conclude that the landscape model, like the object model, is inadequate as a paradigm for the aesthetic appreciation of nature. However, the reason for its inadequacy is instructive. The landscape model is inadequate because it is inappropriate to the nature of the natural environment. Perhaps to see what and how to appreciate in the natural environment, we must consider the nature of that environment more carefully. In this regard there are two rather obvious points that I wish to emphasize. The first is that the natural environment is an environment; the second is that it is natural.

When we conceptualize the natural environment as "nature" I think we are tempted to think of it as an object. When we conceptualize it as "landscape" we are certainly led to thinking of it as scenery. Consequently perhaps the concept of the "natural environment" is somewhat preferable. At least it makes explicit that it is an environment that is under consideration. The object model and the landscape model each in its own way fail to take account of this. But what is involved in taking this into account? Here I initially follow up some remarks made by Sparshott. He suggests that to consider something environmentally is primarily to consider it in regard to the relation of "self to setting," rather than "subject to object" or "traveler to scene."[13] An environment is the setting in which we exist as a "sentient part"; it is our surroundings. Sparshott points out that as our surroundings, our setting, the environment is that which we take for granted, that which we hardly notice – it is necessarily unobtrusive. If any one part of it becomes obtrusive, it is in danger of being seen as an object or a scene, not as our environment. As Sparshott says: "When a man starts talking about 'environmental values' we usually take him to be talking about aesthetic values of a background sort."[14]

The aesthetic values of the environment being primarily background values has obvious ramifications for the questions of what and how to appreciate. Concerning the question of what to appreciate this suggests the answer "everything," for in an essentially unobtrusive setting there seems little basis for including and excluding. I return to this point shortly. Concerning the question of how to appreciate, the answer suggested is in terms of all those ways in which we normally are aware of and experience our surroundings. Sparshott notes that "if environmental aspects are background aspects, eye and ear lose part of their privilege" and goes on to mention smell, touch, and taste, and even warmth and coolness, barometric pressure and humidity as possibly relevant.[15] This points in the right direction, but, as Sparshott also notes, it seems to involve a difficulty – that "the concept of the aesthetic tugs in a different direction" – the direction of the subject/object relation involving primarily the visual scrutiny of an aesthetic object.[16] However, I do not think this difficulty need be as serious as Sparshott seems to think. I suspect the apparent tension here is not due to the concept of the aesthetic being necessarily tied to the subject/object relation or to the visual, but rather is due to its being antithetical to the appreciation of anything only as unobtrusive background. To confirm this we need to consider the concept of the aesthetic as it is elaborated by John Dewey in *Art as Experience*.[17] Dewey's concept is such that anything that is aesthetically appreciated must be obtrusive; it must be foreground, but it need not be an object and it need not be seen (or only seen). Moreover, to assume that that which is aesthetically appreciated need be an object or only seen is to confine aesthetic appreciation to either the object model or the landscape model, which, as we have noted, impose unacceptable limitations on the aesthetic appreciation of the natural environment.

I suggest then that the beginning of an answer to the question of how to aesthetically appreciate an environment is something like the following. We must experience our background setting in all those ways in which we normally experience it, by sight, smell, touch, and whatever. However, we must experience it not as unobtrusive background, but as obtrusive foreground. What is involved in such an "act of aspection" is not completely clear. Dewey gives us an idea in remarks such as:

> To grasp the sources of esthetic experience it is...necessary to have recourse to animal life below the human scale...The live animal is fully present, all there, in all of its actions: in its wary glances, its sharp sniffing, its abrupt cocking of ears. All senses are equally on the *qui vive*.[18]

And perhaps the following description by Yi-Fu Tuan gives some further indication:

> An adult must learn to be yielding and careless like a child if he were to enjoy nature polymorphously. He needs to slip into old clothes so that he could feel free to stretch out on the hay beside the brook and bathe in a meld of physical sensations: the smell of the hay and of horse dung; the warmth of the ground, its hard and soft contours; the warmth of the sun tempered by breeze; the tickling of an ant making its way up the calf of his leg; the play of shifting leaf shadows on his face; the sound of water over the pebbles and boulders, the sound of cicadas and distant traffic. Such an environment might break all the formal rules of euphony and aesthetics, substituting confusion for order, and yet be wholly satisfying.[19]

Tuan's account as to how to appreciate fits well with our earlier answer to the question of what to appreciate, namely everything. This answer, of course, will not do. We cannot appreciate everything; there must be limits and emphases in our aesthetic appreciation of nature as there are in our appreciation of art. Without such limits and emphases our experience of the natural environment would be only "a meld of physical sensations" without any meaning or significance. It would be a Jamesian "blooming, buzzing confusion" that truly substituted "confusion for order" and, I suspect contra to Tuan, would not be wholly satisfying.[20] Such experience would be too far removed from our aesthetic appreciation of art to merit the label "aesthetic" or even the label "appreciation." Consider again the case of art. In this case, as noted in the first section of this chapter, the boundaries and foci of aesthetic significance of works of art are a function of the type of art in question, e.g., paintings end at their frames and their colors are significant. Moreover, I suggested that our knowledge of such matters is due to art works being our creations. Here it is relevant to note the second point that I wish to emphasize about natural environments: they are natural. The natural environment is not a work of art. As such it has no boundaries or foci of aesthetic significance that are given as a result of our creation nor of which we have knowledge because of our involvement in such creation.

The fact that nature is natural – not our creation – does not mean, however, that we must be without knowledge of it. Natural objects are such that we can discover things about them that are independent of any involvement by us in their creation. Thus, although we have not created nature, we yet know a great deal about it. This knowledge, essentially common-sense/scientific knowledge, seems to me the only viable candidate for playing the role concerning the appreciation of nature that our knowledge of types of art, artistic traditions, and the like plays concerning the appreciation of art. Consider the aesthetic appreciation of an environment such as that described by Tuan. We experience the environment as obtrusive foreground – the smell of the hay and of the horse dung, the feel of the ant, the sound of the cicadas and of the distant traffic all force themselves upon us. We experience a "meld of sensations" but, as noted, if our state is to be aesthetic appreciation rather than just the having of raw experience, the meld cannot be simply a "blooming, buzzing confusion." Rather it must be what Dewey called a consummatory experience: one in which knowledge and intelligence transform raw experience by making it determinate, harmonious, and meaningful. For example, in order for there to be aesthetic appreciation we must recognize the smell of the hay and that of the horse dung and perhaps distinguish between them; we must feel the ant at least as an insect rather than as, say, a twitch. Such recognizing and distinguishing results in certain aspects of the obtrusive foreground becoming foci of aesthetic significance. Moreover, they are natural foci appropriate to the particular natural environment we are appreciating. Likewise our knowledge of the environment may yield certain appropriate boundaries or limits to the experience. For example, since we are aesthetically appreciating a certain kind of environment, the sound of cicadas may be appreciated as a proper part of the setting, while the sound of the distant traffic is excluded much as we ignore the coughing in the concert hall.

What I am suggesting is that the question of what to aesthetically appreciate in the natural environment is to be answered in a way analogous to the similar question about art. The difference is that in the case of the natural environment the relevant knowledge is the common-sense/scientific knowledge that we have discovered about the environment in question. This knowledge gives us the appropriate foci of aesthetic significance and the appropriate boundaries of the setting so that our experience becomes one of aesthetic appreciation. If to aesthetically appreciate art we must have knowledge of artistic traditions and styles within those traditions, to aesthetically appreciate nature we must have knowledge of the different environments of nature and of the systems and elements within those environments. In the way in which the art critic and the art historian are well equipped to aesthetically appreciate art, the naturalist and the ecologist are well equipped to aesthetically appreciate nature.[21]

This point about what to appreciate in nature also has ramifications for how to appreciate nature. In discussing the nature of an environment, I suggest that Tuan's description seems to indicate a general act of aspection appropriate for any environment. However, since natural environments differ in type it seems that within this general act of aspection there might be differences which should be noted. To aesthetically appreciate an environment we experience our surroundings as obtrusive foreground, allowing our knowledge of that environment to select certain foci of aesthetic significance and perhaps exclude others, thereby limiting the experience. But certainly there are also different kinds of appropriate acts of aspection which can likewise be selected by our knowledge of environments. Ziff tells us to look for contours in the Florentine school and for color in a Bonnard, to survey a Tintoretto and to scan a Bosch. Consider different natural environments. It seems that we must survey a prairie environment, looking at the subtle contours of the land, feeling the wind blowing across the open space, and smelling the mix of prairie grasses and flowers.

But such an act of aspection has little place in a dense forest environment. Here we must examine and scrutinize, inspecting the detail of the forest floor, listening carefully for the sounds of birds, and smelling carefully for the scent of spruce and pine. Likewise, the description of environmental appreciation given by Tuan, in addition to being a model for environmental acts of aspection in general, is also a description of the act of aspection appropriate for a particular kind of environment – one perhaps best described as pastoral. Different natural environments require different acts of aspection; and as in the case of what to appreciate, our knowledge of the environment in question indicates how to appreciate – that is, indicates the appropriate act of aspection.

The model I thus suggest for the aesthetic appreciation of nature is that which is introduced in Chapter 1 as the natural environmental model. It involves recognizing that nature is an environment and thus a setting within which we exist and that we normally experience with our complete range of senses as our unobtrusive background. But in order for our experience to be aesthetic requires unobtrusive background to be experienced as obtrusive foreground. The result is the experience of a "blooming, buzzing confusion" that in order to be appreciated must be tempered by the knowledge we have discovered about the natural environment so experienced. Our knowledge of the nature of the particular environments yields the appropriate boundaries of appreciation, the particular foci of aesthetic significance, and the relevant act or acts of aspection for that type of environment. We thus have a model that begins to give answers to the questions of what and how to appreciate in the natural environment and that seems to do so with due regard for the nature of that environment. And this is important not only for aesthetic but also for moral and ecological reasons.

Conclusion

In this chapter, in addressing the question of what and how to aesthetically appreciate concerning nature, I contend that two traditional approaches, each of which more or less assimilates the appreciation of nature to the appreciation of certain art forms, leave much to be desired. However, the approach I suggest, the natural environmental model, yet follows closely the general structure of our aesthetic appreciation of art. This approach does not depend on an assimilation of natural objects to art objects nor of landscapes to scenery, but rather on an application of the general structure of aesthetic appreciation of art to something that is not art. What is important is to recognize that nature is an environment and is natural, and to make that recognition central to our aesthetic appreciation. Thereby we will aesthetically appreciate nature for what it is and for the qualities it has. And we will avoid being the person described by Hepburn who "will either pay very little aesthetic heed to natural objects or else heed them in the wrong way," who "will look – and of course look in vain – for what can be found and enjoyed only in art."[22]

Notes

1 Paul Ziff, "Reasons in Art Criticism," *Philosophical Turnings, Essays in Conceptual Appreciation*, Ithaca, Cornell University Press, 1966, p. 71.

2 Ibid. Ziff is mainly concerned with the way in which knowledge of types yields different acts of aspection. For an elaboration of this point and its ramifications concerning what is and is not aesthetically significant in a work, see K. Walton, "Categories of Art," *Philosophical Review*, 1970, vol. 79, pp. 334–67. How our knowledge of art and the artworld yields the boundaries between art and the rest of reality is interestingly discussed in A. Danto, "The Artistic Enfranchisement of Real Objects, the Artworld," *Journal of Philosophy*, 1964, vol. 61, pp. 571–84.

3 F. E. Sparshott, "Figuring the Ground: Notes on Some Theoretical Problems of the Aesthetic Environment," *Journal of Aesthetic Education*, 1972, vol. 6, p. 13.

4 George Santayana, *The Sense of Beauty* [1896], New York, Collier, 1961, p. 100.

5 Danto, op. cit., p. 579. On issues about turning objects into art, see the institutional theory of art; the classic account is G. Dickie, *Art and the Aesthetic, An Institutional Analysis*, Ithaca, Cornell University Press, 1974.

6 This favored sense of "landscape" is brought out by Yi-Fu Tuan. See *Topophilia, A Study of Environmental Perception, Attitudes, and Values*, Englewood Cliffs, Prentice Hall, 1974, pp. 132–3, or "Man and Nature, An Eclectic Reading," *Landscape*, 1966, vol. 15, p. 30.

7 For a good brief discussion of this point, see R. Rees, "The Scenery Cult: Changing Landscape Tastes over Three Centuries," *Landscape*, 1975, vol. 19. Note the following remarks by E. H. Gombrich in "The Renaissance Theory of Art and the Rise of Landscape," *Norm and Form: Studies in the Art of the Renaissance*, London, Phaidon, 1971, pp. 117–18:

> I believe that the idea of natural beauty as an inspiration of art…is, to say the least, a very dangerous oversimplification. Perhaps it even reverses the actual process by which man discovers the beauty of nature. We call a scenery "picturesque"…if it reminds us of paintings we have seen…Similarly, so it seems, the discovery of Alpine scenery does not precede but follows the spread of prints and paintings with mountain panoramas.

8 Thomas West, *Guide to the Lakes* [1778], quoted in J. T. Ogden, "From Spatial to Aesthetic Distance in the Eighteenth Century," *Journal of the History of Ideas*, 1974, vol. 35, pp. 66–7.

9 R. Rees, "The Taste for Mountain Scenery," *History Today*, 1975, vol. 25, p 312.

10 Paul Shepard, *The Tender Carnivore and the Sacred Game*, New York, Scribner, 1973, pp. 147–8. Shepard made this position more explicit at a lecture at Athabasca University, Edmonton, Alberta, November 16, 1974.

11 Rees, "Mountain Scenery," op. cit., p. 312. Ethical worries are also expressed by Tuan, *Topophilia*, op. cit., Chapter 8, and R. A. Smith and C. M. Smith, "Aesthetics and Environmental Education," *Journal of Aesthetic Education*, 1970, vol. 4, pp. 131–2. Smith and Smith put the point as follows:

> Perhaps there is a special form of arrogance in experiencing nature strictly in the categories of art, for the attitude involved here implies an acceptance, though perhaps only momentarily, of the notion that natural elements have been arranged for the sake of the man's aesthetic pleasure. It is possible that this is what Kant had in mind when he said that in the appreciation of natural beauty one ought not assume that nature has fashioned its forms for our delight and that, instead, "it is we who receive nature with favor, and not nature that does us a favor."

12 R. W. Hepburn, "Aesthetic Appreciation of Nature," in H. Osborne (ed.) *Aesthetics and the Modern World*, London, Thames and Hudson, 1968, p. 53. Hepburn argues that our aesthetic appreciation of nature is enhanced by our "realizing" that an object is what it is and has the qualities it has. See pp. 60–5.
13 Sparshott, op. cit., pp. 12–13. Sparshott also considers other possible relations not directly relevant here. Moreover, I suspect he considers the "traveller to scene" relation to be more significant than I do.
14 Ibid., pp. 17–18.
15 Ibid., p. 21.
16 Ibid., pp. 13–14, p. 21.
17 John Dewey, *Art as Experience* [1934], New York, G. P. Putnam's Sons, 1958, especially Chapter 3, "Having an Experience," pp. 35–57.
18 Ibid., pp. 18–19.
19 Tuan, *Topophilia*, op. cit., p. 96.
20 William James, *The Principles of Psychology* [1890], Cambridge, Massachusetts, Harvard University Press, 1983, p. 462.
21 I have in mind here individuals such as John Muir and Aldo Leopold. See, for example, Muir's *Our National Parks*, New York, Houghton Mifflin, 1916, or Leopold's *A Sand County Almanac*, Oxford, Oxford University Press, 1949.
22 Hepburn, op. cit., p. 53.

5

NATURE, AESTHETIC JUDGMENT, AND OBJECTIVITY

Nature and objectivity

In this chapter, I consider the view that some aesthetic judgments about nature and natural objects (e.g., "The Grand Tetons are majestic") are appropriate, correct, or perhaps simply true; while others (e.g., "The Grand Tetons are dumpy") are inappropriate, incorrect, or perhaps simply false. If one accepts some such view, one discovers that the opposition falls into two clearly distinguishable groups. On the one hand, there are those who hold that such views are untenable concerning aesthetic judgments in general, whether a judgment be about nature or about art – whether it be "The Grand Tetons are majestic" or "Donatello's *David* is dumpy." Such objectors often hold some version of a subjectivist, a relativist, and/or a noncognitivist view about all so-called aesthetic judgments. On the other hand, there are those who reject these latter views concerning aesthetic judgments about art, but have serious reservations about rejecting them concerning such judgments about nature. These objectors readily admit and often defend the view that a statement such as, for example, "*Guernica* is dynamic" is appropriate, correct, or true, but find a statement such as "The Grand Tetons are majestic" somewhat worrisome – at least in their theoretical moments, if not in their actual practice. The remarks of this chapter are addressed to only the latter group of objectors. To consider the issues raised by the former group would require not only a much more substantial investigation, but also one with an essentially different focus.

Although the position of the latter group is quite common among aestheticians, it can be properly discussed only by considering a concrete example. One such example is found in Kendall Walton's important essay "Categories of Art."[1] I select Walton's position for consideration for two reasons. First, it presents a persuasive and well-developed account of the truth and falsity of aesthetic judgments about art, an account that does not directly apply to aesthetic judgments about nature, and thus appears to necessitate a relativist view in regard to these judgments. Second, the position is developed in such a way that it makes possible a clear understanding of why aestheticians might hold an essentially objectivist view concerning aesthetic judgments about

Illustration 2 The Grand Tetons, Wyoming (*Courtesy of Corbis/Ansel Adams Trust*).

art and yet a relativist view concerning those about nature. My general point in considering Walton's position is to demonstrate that a somewhat analogous position does in fact apply to aesthetic judgments about nature. If this is the case, then to the extent that positions of this general type underwrite the objectivity of aesthetic judgments about art, the objectivity of aesthetic judgments about nature is similarly underwritten.

Walton's position

To begin we must have at our disposal a brief and schematic account of Walton's position together with his reservations concerning its applicability to aesthetic judgments about nature.[2] Essentially Walton holds that the truth value of aesthetic judgments about a work of art is a function of two things: first, the (nonaesthetic) perceptual properties a work actually has, and second, the perceived status of such perceptual properties when a work is perceived in its *correct* category or categories of art. He provides evidence for the psychological claim that the aesthetic judgments that *seem* true or false of a work are a function of the perceived status of its perceptual properties, given any category in which the work is perceived; he argues for the philosophical claim that the aesthetic judgments that *are* true or false of a work are a function of the perceived status of its perceptual properties, given that the work is perceived in its correct category or categories.[3] He then provides four circumstances which count toward the correctness of perceiving a work in a category (see below).

The above sketch is most succinctly elaborated by means of an example. Consider "*Guernica* is awkward." If this aesthetic judgment is false, it is false as a function of the perceptual properties of *Guernica* (1937), that is, its lines, colors, forms. With this few would disagree. Walton goes further to point out that *Guernica* can be perceived in different (perceptually distinguishable) categories of art.[4] It can be perceived, for example, as a painting, as an impressionist painting, or as a cubist painting. With respect to such categories of art, certain perceptual properties are what Walton calls standard, contra-standard, and variable.[5] For example, flatness is standard for all three of the above-mentioned categories, being colored is variable for all three, but having predominately cube-like shapes is variable for the first, contra-standard for the second, and standard for the third. The perceived status of perceptual properties as standard, variable, or contra-standard is thus a function of the category in which the work is perceived.[6] If *Guernica* is perceived as a cubist painting, its cube-like shapes will be perceived as standard; if it is perceived as an impressionist painting these same shapes will be perceived as contra-standard (or possibly as variable). Walton's psychological claim is that this perceived status affects which aesthetic judgments seem true or false to a perceiver on an occasion. "*Guernica* is awkward" may seem a true judgment when the painting is seen in the category of impressionist paintings, for its cube-like shapes will be perceived as contra-standard (or variable) and thus as counting toward awkwardness. The same judgment will seem false, however, when the painting is seen in the category of cubist paintings, for then its cube-like shapes will be perceived as standard and consequently as counting toward awkwardness no more than its flatness as a painting, for example, is perceived as counting toward its representing flat objects.[7]

But is "*Guernica* is awkward" really true or really false? Walton's philosophical claim is that this depends on the perceived status of the perceptual properties when *Guernica* is perceived in its *correct* category. In this particular case the aesthetic judgment in question is false because *Guernica* is correctly perceived as a cubist painting; thus the perceived status of its cube-like shapes does not count toward awkwardness. The four circumstances that count toward it being correct to perceive *Guernica* as a cubist painting are:

- that it has a relatively large number of properties standard with respect to cubism;
- that it is a better painting when perceived as a cubist painting;
- that Picasso intended or expected it to be perceived as a cubist painting;
- that the category of cubist paintings was well established in and recognized by the society in which *Guernica* was produced.

Walton argues that generalized versions of these four circumstances are relevant to determining the correct category within which to perceive any work of art.[8]

It follows from Walton's account that, in order to determine the truth value of an aesthetic judgment such as "*Guernica* is awkward," it will not do simply to look at *Guernica*, as it will if we wish to determine the truth value of "*Guernica* is colored." Rather, we must perceive *Guernica* in its correct category. This requires two kinds of knowledge. First, the knowledge *that* certain factors make cubism its correct category and consequently that cubism is its correct category; that is, certain factual knowledge about the history and nature of twentieth-century art. And second, the knowledge *how* to perceive *Guernica* as a cubist work; that is, certain practical knowledge or skill that must be acquired by training and experience concerning the category of cubist paintings and other related categories of art. In short, simply "examining a work with the senses can by itself reveal neither how it is correct to perceive it, nor how to perceive it that way."[9]

We are now in a position to consider the relevance of a Walton-like position to aesthetic judgments about nature. Walton holds that his philosophical claim is not applicable to most aesthetic judgments about nature, and that these judgments are perhaps better understood in terms of what he calls the category-relative interpretation. On this interpretation aesthetic judgments are relativized to any category in which a person happens to perceive something. For example, if this interpretation were applied to aesthetic judgments about works of art, then judgments such as "*Guernica* is awkward" and *Guernica* is not awkward" need not be incompatible, for the first may amount to the judgment that "*Guernica* is awkward as an impressionist painting," and the second to the judgment that "*Guernica* is not awkward as a cubist painting." Walton argues that, in regard to aesthetic judgments about art, this interpretation is inadequate for it does not allow such aesthetic judgments to be mistaken often enough. We say that the judgment "*Guernica* is awkward" is simply wrong, false; we say that the individual making such a judgment does not appreciate the work because he or she perceives it incorrectly. Similarly, we say that judgments we ourselves made about works before we appreciated them properly were simply false or wrong. But the category-relative interpretation does not allow for these facts; thus, in order to accommodate them we must accept that certain ways of perceiving works of art are correct and others incorrect.

However, although Walton rejects the category-relative interpretation of judgments about art, he proposes it for aesthetic judgments about nature. He says: "I think that aesthetic judgments are in *some* contexts amenable to such category-relative interpretations, especially aesthetic judgments about natural objects (clouds, mountains, sunsets) rather than works of art."[10] He suggests that we simply *attribute* aesthetic properties to natural objects in the same way in which we could, or would have to, concerning a work of art "about whose origins we know absolutely nothing" and that we are, consequently, *not* "in a position to *judge...aesthetically*."[11]

Nature and culture

It should be clear that Walton's account is a concrete example of one of the positions mentioned at the beginning of this chapter: aesthetic judgments about art are true or false and can be determined as such, but aesthetic judgments about nature are in some sense subjective or relative. Walton has an essentially objectivist account of aesthetic judgments about art and yet a relativist account of those about nature. Concerning the latter, the implicit view is that such judgments are relative to the way in which a perceiver happens to perceive a part of nature or a natural object on a particular occasion. There seem to be no essentially correct or incorrect categories in which to perceive nature, and thus with nature one can only *attribute* to it certain aesthetic properties rather than *judge* which aesthetic properties it has. With nature it appears to be a matter of aesthetically appreciating whatever one can and as much as one can, but *not* a matter of getting it aesthetically right or wrong, *not* a matter of making true or false aesthetic judgments.

This position has an initial implausibility that leads one to ask why Walton and other aestheticians accept it. Its implausibility can be seen by noting that not only do many aesthetic judgments about nature strike us as clearly true (e.g., "The Grand Tetons are majestic") or clearly false (e.g., "The Grand Tetons are dumpy"), but also that many of such judgments seem to be paradigmatic aesthetic judgments – ones in virtue of which we initially grasp aesthetic concepts (e.g., graceful gazelle, majestic mountain, sublime sunset). In light of these considerations a position that bifurcates the class of aesthetic judgments and suggests an essentially different and weaker philosophical account of those about nature than of those about art seems counterintuitive. If such an account is inadequate, it might be suggested that this is due to the fact that, as explained in Chapter 1, aestheticians have until very recently paid relatively little attention to the aesthetics of nature.[12] However, although it is perhaps true that philosophers hold less adequate views about that to which they give less attention, this cannot be the complete explanation for a position such as that sketched above. There are more important reasons that can be made explicit by further considering our example.

As we saw above, Walton provides four circumstances that count toward it being correct to perceive a work in a given category of art. The latter two circumstances are respectively that the artist who produced the work intended or expected it to be perceived in a certain category or thought of it as being in that category, and that the category in question is well established in and recognized by the society in which the work is produced. These circumstances cannot count toward it being correct to perceive parts of nature or natural objects in certain categories, for nature is not produced by artists who intend or expect it to be perceived in certain ways and it is not produced *within* certain societies. Consequently an important aspect of Walton's philosophical account is, as Walton says, "not readily applicable to most judgments about natural objects."[13] Essentially, this is because works of art are produced by artists within societies and nature is not. Moreover, not only are works of art the products of societies, but in a similar way so are the categories of art themselves. Consequently, not only do the circumstances for a category being correct not readily apply to nature, but the categories themselves do not apply. In short, nature does not fit into categories of art.

These remarks enable us to see more clearly the basic reason why some aestheticians hold an essentially objectivist view concerning aesthetic judgments about art and a relativist view concerning those about nature. In short, it is because the objectivity of the former class of aesthetic judgments is based on a certain kind of account of the nature of aesthetic appreciation, aesthetic qualities, and/or aesthetic objects. The relevant accounts are in some sense societal, institutional, or more generally, cultural.[14] The two circumstances of Walton mentioned above make reference to certain cultural facts and his categories are themselves culturally established and maintained. It is the difficulty of bringing aesthetic judgments about nature into the general framework of these cultural accounts that seemingly leads aestheticians to abandon such judgments to relativism.[15]

Once it is clear that cultural accounts of the aesthetic are the basic reason for a bifurcated view of aesthetic judgments, we might be tempted to remedy the situation by rejecting such accounts. However, although there are philosophical problems with these accounts, I think this temptation should be resisted.[16] Cultural accounts of the aesthetic are simply more promising than other alternatives. Consequently, in regard to the aesthetics of nature, the interesting issue is the extent to which we can develop an analogous account which applies to aesthetic judgments about nature. In order to pursue this issue we must consider carefully the important differences between art and nature that cultural accounts of the aesthetic force to our attention. I have briefly indicated these differences in this section, but they need to be further discussed if we are to move toward a unified account of aesthetic judgments. I return to them shortly. Initially, however, it is fruitful to see exactly how much of a cultural account roughly of the kind outlined by Walton does directly apply to nature and how well it applies. This will allow us to see more clearly exactly what and how much needs to be said about these differences in order to achieve a unified account.

Nature and Walton's psychological claim

Walton's position embodies not only a philosophical claim but also a psychological claim. The latter is the claim that the aesthetic judgments that *seem* true or false of a work are a function of the perceptual status of its perceptual properties given any category in which the work is perceived. I think that a similar psychological claim can be demonstrated to hold for aesthetic judgments about nature. In fact Walton presents one case which suggests that this is so:

> A small elephant, one which is smaller than most elephants with which we are familiar, might impress us as charming, cute, delicate, or puny. This is not simply because of its (absolute) size, but because it is small *for an elephant.* To people who are familiar not with our elephants but with a race of mini-elephants, the same animal may look massive, strong, dominant, threatening, lumbering, if it is large for a mini-elephant. The size of elephants is variable relative to the class of elephants, but it varies only within a certain (not precisely specifiable) range. It is a standard property of elephants that they do fall within this range. How an elephant's size affects us aesthetically depends, since we see it as an elephant, on whether it falls in the upper, middle or lower part of the range.[17]

This case illustrates the application of the psychological claim to natural objects and provides an example of a relevant category with properties standard, variable, and contra-standard in respect to that category. The category is that of elephants, a perceptually distinguishable category in which we do perceive certain natural objects – under most conditions, elephants. With respect to this category, size is a variable property while the limits of range on size is a standard property. Thus the small elephant's size is variable in this example. However, were the elephant small enough to fall outside the limits set by the standard property – for example, the size of a large mouse – its size would be contra-standard. In such a case, if we could still perceive it in the category of elephants, other aesthetic judgments would seem true. In addition to (or as opposed to) appearing delicate, it might appear frail or fragile. And it would certainly strike us as being surprising or disconcerting – aesthetic properties related to certain contra-standard perceptual properties. The point illustrated by this case generally holds true for similar cases. Consider, for example, the aesthetic judgments we take to be true of Shetland ponies (charming, cute) and Clydesdale horses (majestic, lumbering). These judgments are made with respect to the category of horses. Similarly a foal (calf, fawn, etc.) typically strikes us as delicate and nimble when seen in the category of horses (cattle, deer, etc.), but a particularly husky one may strike us as lumbering or perhaps awkward if seen in the category of foals (calves, fawns, etc.).

In the above example, particular natural kinds (elephants, horses) constitute categories that function psychologically as do categories of art; and our aesthetic appreciation is directed toward a natural *object.* We also need, however, to consider the application of the psychological claim to a somewhat different kind of example – the aesthetic appreciation of landscapes or of natural environments. In discussing the enriching of aesthetic appreciation, Ronald Hepburn describes the following case:

> Supposing I am walking over a wide expanse of sand and mud. The quality of the scene is perhaps that of wild, glad emptiness. But suppose

> that I bring to bear upon the scene my knowledge that this is a tidal basin, the tide being out. I see myself now as virtually walking on what is for half the day sea-bed. The wild, glad emptiness may be tempered by a disturbing weirdness.[18]

This case may be elaborated as follows. Note that what is described is a change in which aesthetic judgments seem true of the "wide expanse of sand and mud," and this change is a function of perceiving the expanse in different ways. Initially it is apparently perceived as a beach, but then due to the realization that it is half the day under the sea, it is perceived as a sea-bed. Here beach and sea-bed – along the tidal basin – function as categories.[19] These categories are perceptually distinguishable in terms of their perceptual properties, and we can and do perceive things such as a "wide expanse of sand and mud" as in or as belonging to them. Moreover, such perceiving is not a matter of inferring, but rather a matter of simply seeing the expanse in the relevant categories given our experience and knowledge. It is a matter of perceiving a number of perceptual properties characteristic of (standard for) these categories combined into a single Gestalt (see footnote 6).

Once we recognize beach and sea-bed as categories, the aesthetic judgments that Hepburn mentions can be accounted for in terms of standard, variable, and contra-standard properties in respect to such categories. Perceiving the expanse in the category of sea-beds (as opposed to beaches) results in its appearing to possess a "disturbing weirdness." The property of being relatively dry (or above water) and thus such that it can be walked upon is a standard property with respect to beaches and contra-standard with respect to sea-beds. Dryness is among those properties in virtue of which such an expanse belongs to the category of beaches and which tends to disqualify it from belonging to the category of sea-beds (see footnote 5). Thus perceiving the expanse in the category of sea-beds results in its dryness (and the walking upon it) being perceived as contra-standard. Contra-standard properties are ones we tend to find "shocking, disconcerting, startling, or upsetting," and thus results the "disturbing weirdness." The weirdness is the result not simply of the realization that one is walking on a tidal basin, but of the experience of walking where it is, as it were, contra-standard to walk, that is, in perceiving the wide expanse as a sea-bed and perceiving oneself as walking upon that sea-bed.

Similarly, concerning properties that are standard, variable, and contra-standard in respect to beaches we might explain the initial aesthetic judgment involving "wild, glad emptiness." As in the case of the size of elephants, the width and expansiveness of, in this case, a beach is a variable property and the limits on the range of width and expansiveness a standard property with respect to this category. Thus if the "sand and mud" is, as suggested by the quote, quite wide and expansive, that is, in the upper part of the range, the resultant aesthetic impression will be of a "wild, glad emptiness." On the other hand, had the expanse of sand and mud been in the lower part of the range, the beach may have appeared warm and cozy or perhaps cramped and confining. Of course, the actual aesthetic judgments that seem true will be a function not only of these variable and standard properties of beaches, but also of many others and, of course, of our other experiences with this category and the properties that are standard, variable, and contra-standard with respect to it.

Other examples similar to the above could be provided. In fact I think reference to the kind of categories in which we perceive natural objects and landscapes, and the properties of standard, variable, and contra-standard with respect to such categories, helps to explain which aesthetic judgments seem to be true or false concerning much of nature – whether it be completely natural or modified by humans. Reflection on our judgments about mountains, sunsets, and waving fields of grain appears to bear this out.[20] However, rather than pursue further examples that support this psychological claim, I now turn to the more important question of the correctness of such judgments. In short, given the truth of the psychological claim, I ask whether we should, with Walton and others, opt for the category-relative interpretation of our aesthetic judgments about nature; or alternatively whether there are persuasive arguments for considering certain categories of nature to be correct and others incorrect, as is the case with categories of art. Given the implausibility of a bifurcated account of aesthetic judgments, I assume that the existence of such arguments constitutes adequate reason for rejecting the category-relative interpretation.

The correct categories of nature

In the remainder of this chapter I sketch some arguments that give grounds for holding that certain categories of nature are correct and others not. These arguments support the conclusion suggested in the preceding chapter: that the correct categories of nature are those given by the natural sciences. Before considering these arguments, however, we should note that there is one obvious way in which certain of such categories are correct and others not. It appears, for example, that the category of elephants and not that of mice is the correct category for perceiving an elephant, regardless of its size. Similarly, to perceive an elephant as a mountain, as a sunset, or as a waving field of grain are clear cases of getting the incorrect category – if such perceivings are even possible. These are easy cases, for in such cases the number of perceptual properties standard with respect to the correct category are relatively large in number – even a very small (or very large) elephant has considerably more perceptual properties standard with respect to the category of elephants than with respect to the category of mice (or mountains). In short, in such cases the correct category is determined by something like Walton's first circumstance for it being correct to perceive a work of art in a given category – and determined by that alone. The more difficult kind of case is that in which the perceptual properties of a natural object or a part of nature do not by themselves clearly indicate a correct category. Such a case is posed when, given the perceptual properties, it is yet plausible to perceive an object in two or more mutually exclusive categories. Very simple examples are perceiving a sea-anemone as a plant or as an animal, or perceiving a whale as a fish or as a mammal. Yet, even with these examples, it may be argued that perceptual properties are adequate to determine correctness of category – depending, of course, on what properties count as perceptual.

In order to consider my first argument, however, it is not necessary to debate the difficult issue of what counts as a perceptual property. Rather it is more fruitful to simply assume that in such cases perceptual properties are not adequate to determine correctness of category. Consequently, let us assume, for example, that with perceptual properties alone it is plausible to perceive a whale either in the category of fish or in the category of mammals. This makes plausible the claim that for aesthetic judgments about whales there is no correct category, and that we must therefore accept the category-relative interpretation of such judgments. We can see why anyone with a Walton-like position might be moved to this view. Given our assumption, Walton's first circumstance does not determine a correct category and, as already suggested, his third and fourth circumstances cannot determine a correct category.[21] Whales are not produced by artists who intend them to be perceived in certain categories and are not produced within societies. Whales do not fit into categories of art. However, from these truisms, Walton and others apparently move to the view that concerning nature we are completely without resources for determining a correct category. As we have seen, Walton compares making aesthetic judgments about nature with making aesthetic judgments about a work of art "about whose origins *we know absolutely nothing*."[22]

It is this move and the resulting conclusion that my first argument calls into question. To see how this occurs we must note first that although whales and nature in general do not fit into categories of art, they yet do fit, as noted in Chapter 4, into a number of common-sense and/or "scientific" groups. A whale can be perceived as a fish or as a mammal; as a whale or as, for example, a large porpoise; as a blue whale or as a humpbacked whale. Hepburn's expanse of sand and mud can be perceived as a beach, as a tidal basin, or as a sea-bed. That nature in general can be and is perceived in such biological and geological categories and that such categories psychologically function in a way similar to categories of art is what has already been established by the above discussion of the application of the psychological claim to aesthetic judgments about nature. Once this is granted, we must next observe that although nature is not produced by artists who intend it to be perceived in certain categories and is not produced within certain societies, it does not follow that we know nothing about it or more particularly that we do not know which categories are correct for it. Works of art are *produced* within societies within which categories are recognized and by artists who intend these to be correct categories; it follows that certain categories are correct for a given work and that we can know which ones are correct. But, as suggested in Chapter 4, human production is not the only key to correctness of category. In general we do not produce, but rather discover, natural objects and aspects of nature. Why should we therefore not discover the correct categories for their perception? We discover whales and later discover that, in spite of somewhat misleading perceptual properties, they are in fact mammals and not fish.[23] It is plausible to claim that we have discovered the correct category in which to perceive whales. In the first place, that whales are not of our production does not count *against* this category being correct. In the second, it fits our intuitions about the correct categorization or classification of whales if this issue is considered independent of aesthetic issues. And, in the third place, this correctness of category can function philosophically for aesthetic judgments about nature as the correctness of categories of art functions for aesthetic judgments about art works. The only significant difference is that, concerning the latter, the grounds for correctness are the activities of artists and art critics, while concerning the former, they are the activities of naturalists and scientists in the broadest sense. But given the differences between art and nature, this is only to be expected.

The above line of thought takes us only so far in undercutting the plausibility of the category-relative interpretation of aesthetic judgments about nature. One might say that it at best shifts the burden of proof to those who defend such an interpretation. Moreover, it is based on the (at best unclear) assumption that in a case such as that of whales perceptual properties themselves do not determine the correct category. Consequently we must consider some further arguments that do more than shift the burden of proof and do not depend upon this assumption. To find a realistic case in which perceptual properties are neutral concerning correctness of category it is useful to envisage a situation where humans are involved. We, or landscapers, sometimes construct (or reconstruct) landscapes that are perceptually indistinguishable from natural landscapes. Consider a scenic coastline that appears to be natural, but in fact has been created by human beings. Imagine that it has been carefully planned and designed to be perceptually indistinguishable from a natural coastline, but its construction involved the removal of buildings and parking lots, the redistribution of great quantities of sand and soil, and the landscaping of the whole area to blend with its surroundings.[24] It is in fact a large-scale artifact, but can, and probably would, be perceived in the category of apparent (see footnote 19) natural coastlines. In such a case perceptual properties alone clearly do not *determine* whether it is correct to perceive the landscape as natural coastline or as artifact, yet the question of which is the correct category can be raised. It is essentially the question of whether it is correct to perceive the object as what it is (an artifact) or as what it appears to be (a natural coastline). It is this form of the question to which I address two additional lines of thought to the effect that certain categories in which we perceive nature are correct and others not. The general direction of these arguments is that it is correct to perceive an object in the category of what it is (as opposed to what it appears to be) even in difficult cases where its perceptual properties do not themselves count toward one or the other categories being correct (or, if anything, suggest that the category of what it appears to be is correct).

The first of these two additional arguments can be made clear by imagining another coastline. This is a coastline that is perceptually indistinguishable from the above described coastline, but is natural rather than human-made. We can assume it is the coastline that served as the model for the human-made one. Since the two coastlines (call them N for natural and M for human-made) are perceptually indistinguishable, there is one level at which our aesthetic appreciation of the two will be identical. This is the level at which we appreciate only perceptual properties such as the curves, lines, colors, shapes, and patterns of N and M. However, as argued in Chapters 2 and 3, such exclusively formal aesthetic appreciation of nature is problematic.[25] Moreover, as suggested in Chapter 4, there is also another, deeper level at which aesthetic appreciation occurs.[26] At this level we appreciate not simply, for example, the identical patterns of N and M, but such patterns under certain descriptions. For example, the pattern of M can be described as indicating careful design, as an exact copy of the pattern of N, or as the product of human ingenuity; while the pattern of N can be described as typical of, say, North American Pacific coastlines, as indicating a high-tide coastal formation, or as the product of the erosion of the sea. It is clear that aesthetic appreciation of perceptual properties under such descriptions constitutes an important part of aesthetic appreciation of nature.[27] It essentially involves the contemplation of perceptual properties in light of such descriptions and the appreciation of them as something in virtue of which such descriptions are true.

Given the aesthetic relevance of these kinds of descriptions, it must be recognized that the above descriptions of the patterns of N and M are such that those that are true of N's pattern are not true of M's and vice versa. In fact for any such N and M there are an infinite number of descriptions that are true of N and not of M and vice versa. That any of such descriptions are aesthetically relevant in the way suggested above is all that need be the case in order to establish the importance of perceiving an object in the category of what it *is* as opposed to what it *appears to be*. This can be seen by noting that if we, for example, perceive M in the category of natural coastlines (what it appears to be) we become involved in one or both of the following: first, failure to appreciate it under descriptions that are true of it, such as its being carefully designed by humans; second, appreciation of it under descriptions that are false of it, such as its being the result of the sea's erosion. The first alternative is undesirable as it constitutes a case of aesthetic omission. It is possible to contemplate M in light of the description "being carefully designed by humans" and moreover to appreciate it as something in virtue of which this is true, but these possibilities are not likely to be achieved if M is perceived (only) in the category of natural coastlines. The opportunity for such contemplation and appreciation is not provided by perceiving M in this category. The second alternative is undesirable as it constitutes a case of aesthetic deception.

Perceiving M in the category of natural coastlines, of course, provides the possibility of contemplating M in light of the description "being the result of the sea's erosion," but this is a tenuous and misleading contemplation. It is tenuous in that it is always in danger of being destroyed by the knowledge of M's true reality and it is misleading in that it directs our contemplation away from this reality. Moreover, if we appreciate M as something in virtue of which this description is true, we are simply mistaken; our appreciation involves a false belief. On the other hand, if we perceive M in the category of artifact or perceive N in the category of natural coastlines – the categories of what each in fact is – then aesthetic omissions and aesthetic deceptions of the kind described above need not occur.

There are two things to observe about the preceding argument. The first and most important is that it provides grounds for construing the category of what something is as the correct category and the category of what something only appears to be as an incorrect category. The grounds are essentially that doing so avoids both aesthetic omissions and aesthetic deceptions. Consequently, we have reason to hold that even in cases where perceptual properties do not by themselves determine correct and incorrect categories, there are yet grounds for this determination and, therefore, grounds for rejecting the category-relative interpretation of our aesthetic judgments about nature. The second is that this conclusion depends neither upon the example involving very broad categories (natural coastline versus artifact or human-made coastline) nor upon it involving one "non-natural" category. As suggested earlier, these features of the example are due to the desire for a realistic case in which perceptual properties themselves cannot be construed as determining the correctness of one or the other category. In this regard it is important to note that this same line of argument applies to, for example, our earlier case involving the categories of beach, sea-bed, and tidal basin. Different aesthetically relevant descriptions are true of the wide expanse of sand and mud depending upon whether it is in fact a beach, a sea-bed, or a tidal basin. And only perceiving the expanse in the category of what it is will avoid various aesthetic omissions and deceptions. Consequently, in such cases, we similarly have grounds for the determination of the correct category and for rejecting the category-relative interpretation of aesthetic judgments.

The second argument for construing the category of what a natural object is as the correct category and what it only appears to be as an incorrect category is in part an ethical argument. It is essentially the contention that this is the best way to keep our aesthetics and our ethics in harmony. Consider the aesthetic appreciation of a *Playboy* centerfold model.[28] Whether or not this is considered aesthetic appreciation of nature, there is a common line of argument concerning such aesthetic appreciation. It is argued that this is to aesthetically appreciate the model not as what she is (in the category of human beings), but only as what she here appears to be or is presented as being (in the category of sex objects). And, the argument continues, this is ethically suspect for to engage in such aesthetic appreciation is to endorse and promote (in ourselves, if nowhere else) a sexist attitude toward woman. I think this *kind* of argument has merit for the following reason: it is clear that we do not aesthetically appreciate simply with our five senses, but rather with an important part of our whole emotional and psychological selves. Consequently, what and how we aesthetically appreciate cannot but play a role in the shaping of our emotional and psychological being. This in turn helps to determine what we think and do, and think it correct for ourselves and others to think and do. In short, our aesthetic appreciation is a significant factor in shaping and forming our ethical views.

If this argument has merit, it is especially pertinent to the aesthetic appreciation of nature. If our aesthetic appreciation of nature helps to determine our ethical views concerning nature, then our aesthetic appreciation of nature should be of nature as it in fact is rather than as what it may appear to be. By aesthetically appreciating nature for what it is, we will shape our ethical views such that there is the best opportunity for making sound ethical judgments about matters of environmental and ecological concern. Consider again the human-made coastline. What if we discover that it causes environmental and ethical problems? Perhaps it greatly decreases the possibility of successful upstream migration by spawning salmon, or perhaps it causes an undercurrent that is exceedingly dangerous to swimmers. If we perceive the coastline in the category of natural coastlines (and are entrenched in doing so), a sound ethical view might involve noting that fish and human beings have in such cases long accepted and met the challenges of nature. Consequently perhaps we understandably conclude that we should let nature take its course and swimmers take their chances. On the other hand, if we perceive the coastline in the category of artifact or human-made coastline, a sound ethical view might involve regarding our environmental and ethical responsibilities quite differently. Perhaps we, ethically and ecologically, should construct a fish ladder up the coast (as has been done to allow salmon migration around hydroelectric dams), and perhaps we, ethically, should forbid swimmers to use the area. An actual case of this kind might be the way in which ethical views about whales apparently alter as a function of perceiving them as mammals rather than as fish. At the very least it appears that *some* of the arguments advanced for preserving whales presuppose perceiving them as mammals and aesthetically appreciating them as such. Strictly analogous arguments could not be advanced for preserving, for example, sharks (although different arguments could be).

This ethical line of argument does not by itself clearly establish that there are correct and incorrect categories in which to perceive parts of nature or natural objects nor does it clearly establish that the correct categories are the categories of what things in fact are. However, it does, I think, establish that there is ethical merit in regarding certain categories as correct and others as incorrect, and in regarding as correct the categories of what things in fact are, or, as we described them earlier, the common-sense and/or scientific categories that are determined by the naturalist and the natural scientist. Consequently, if this argument together with the others offered above give adequate grounds for claiming truth and falsity for our aesthetic judgments of nature (rather than accepting the category-relative interpretation), then these arguments help to establish a position that has additional merit. This is the merit of bringing the interests and points of view of aesthetics, ethics, and natural science together such that they reinforce one another, rather than stand in opposition as they so often appear to do.

Conclusion

I conclude these remarks by further emphasizing one consequence of the position I have attempted to establish. The consequence can be brought out by noting the extent to which the position remains within the general confines of a cultural account of the aesthetic. There is a difference of emphasis, of course. As noted in Chapter 4, when the aesthetic appreciation of art is considered in light of a cultural account, the relevant part of our culture is that embodied in and revealed by art history and art criticism; when aesthetic appreciation of nature is so considered, on the other hand, the relevant part of our culture is natural history and natural science. Nonetheless, in the manner in which a cultural account of the aesthetic requires knowledge of art history and art criticism to play an essential role in our aesthetic judgments about art, likewise a cultural account requires knowledge of natural history and natural science to play the same essential role in our aesthetic judgments about nature. We can, of course, approach nature as we sometimes approach art, that is, we can simply *enjoy* its forms and colors or *enjoy* perceiving it however we may happen to. But if our appreciation is to be at a deeper level, if we are to make aesthetic judgments that are likely to be true and to be able to determine whether or not they are true; then we must know something about that which we appreciate. We must know *that* certain factors make aspects of nature and natural objects belong to certain categories and that they are therefore correctly perceived in these categories. And we must know *how* to perceive those aspects of nature and natural objects in the categories in question.[29] This reinforces the conclusion reached in Chapter 4: that for significant aesthetic appreciation of nature, something like the knowledge and experience of the naturalist is essential. It is not surprising that individuals such as Muir, Ruskin, Audubon, and Leopold, who demonstrated an acute aesthetic appreciation of nature in their paintings and writings, were not simply appreciators of nature but also accomplished naturalists.

Notes

1 Kendall Walton, "Categories of Art," *Philosophical Review*, 1970, vol. 79, pp. 334–67.

2 The following necessarily brief sketch of Walton's position fails to capture its detail and subtlety. Footnotes 4 to 7 are designed to supplement the sketch to some extent.

3 Walton puts his psychological and his philosophical claims in terms of what aesthetic properties a work seems to have or has, as opposed to in terms of what aesthetic judgements appear to be or are true or false of it. I prefer the latter and nothing, I believe, turns on this way of putting his position.

4 Walton, op. cit., pp. 338–9:

> Such categories include media, genre, styles, forms, and so forth – for example, the categories of paintings, cubist paintings, Gothic architecture, classical sonatas, paintings in the style of Cezanne, and music in the style of late Beethoven – if they are interpreted in such a way that membership is determined solely by features that can be perceived in a work when it is experienced in the normal manner.

The latter condition is what makes a category of art "perceptually distinguishable."

5 Ibid., p. 339:

> A feature of a work of art is *standard* with respect to a (perceptually distinguishable) category just in case it is among those in virtue of which works in that category belong to that category – that is, just in case the lack of that feature would disqualify, or tend to disqualify, a work from that category. A feature is *variable* with respect to a category just in case it has nothing to do with works belonging to that category; the possession or lack of the feature is irrelevant to whether a work qualifies for the category. Finally, a *contra-standard* feature with respect to a category is the absence of a standard feature with respect to that category – that is, a feature whose presence tends to *disqualify* works as members of the category.

6 Ibid., p. 340

> To perceive a work in a certain category is to perceive the "*Gestalt*" of that category in the work. This needs some explanation. People familiar with Brahmsian music – that is, music in the style of Brahms (notably, works of Johannes Brahms) – or impressionist paintings can frequently recognize members of these categories by recognizing the Brahmsian or impressionist *Gestalt* qualities. Such recognition is dependent on perception of particular features that are standard relative to these categories, but it is not a matter of *inferring* from the presence of such features that a work is Brahmsian or impressionist.

As is evident in the above and in footnote 4, Walton uses "perceive" in a rather broad sense. I follow his usage throughout this chapter.

7 In general Walton's view is that properties perceived as standard are aesthetically inert or "contribute to a work's sense of order, inevitability, stability, correctness" Ibid., p. 348; properties perceived as variable contribute to a work its representational, symbolic, and expressive nature; and properties perceived as contra-standard contribute to a work a shocking, disconcerting, startling, or upsetting nature. Ibid., see pp. 343–54.

8 Ibid., see pp. 357–63.

9 Ibid., p. 367.

10 Ibid., p. 355.

11 Ibid., p. 365, my italics.

12 This has been noted in different ways by a number of writers. See, for example, Ronald W. Hepburn, "Aesthetic Appreciation of Nature," in H. Osborne (ed.) *Aesthetics in the Modern World*, London, Thames and Hudson, 1968, pp. 49–66; Mary Carman

Rose, "Nature as an Aesthetic Concept," *British Journal of Aesthetics* 1976, vol. 16, pp.3–12; Michael Hancher, "Poems Versus Trees: The Aesthetics of Monroe Beardsley," *Journal of Aesthetics and Art Criticism*, 1972, vol. 31, pp. 181–91.

13 Walton, op. cit., p. 355.

14 It should be clear that the position concerning the aesthetics of nature that I here exemplify by means of Walton's article can also be seen as a natural consequence of George Dickie's institutional analysis of aesthetic objects (not simply of art). See *Art and the Aesthetic: An Institutional Analysis*, Ithaca, Cornell University Press, 1974, Chapters 7 and 8, especially p. 169 and pp. 198–200. I do not consider Dickie's analysis as he does not explicitly endorse a relativist position for aesthetic judgments about nature. He does however make certain remarks that suggest it. For example: "Where natural visual aesthetic objects are concerned, we can be content to appreciate whatever happens to 'fall together' into a visual design" (Ibid., p. 169). In spite of such comments, it is nonetheless clear from the concluding paragraphs of the book that Dickie wishes to leave open the question of the status of the aesthetic appreciation of nature. The view that I elaborate in this chapter constitutes a partial response to some of the issues raised in those paragraphs.

15 As noted in Chapters 1 and 4, one means by which aesthetic judgments about nature can be brought under cultural accounts of the aesthetic is in virtue of the practice of perceiving nature *as if* it fits into certain kinds of *artistic* categories. There is, for example, a tradition of perceiving landscapes as if they fit into various landscape-painting categories. Paradoxically, however, such perceiving of landscapes seemingly lends support to a category-relative interpretation of aesthetic judgments of nature, for it is difficult to justify the claim that any such artistic categories are *correct for* nature, rather than simply *imposed upon* nature. I think imposing such categories upon nature poses certain difficulties that I discuss in "Appreciation and the Natural Environment," *Journal of Aesthetics and Art Criticism*, 1979, vol. 37, pp. 267–75 (reproduced in this volume, Chapter 4).

16 There is, of course, no lack of philosophical problems with cultural accounts of the aesthetic. Some of the particular problems inherent in Dickie's institutional analysis of aesthetic objects are brought out in the following: Gary Iseminger, "Appreciation, the Artworld, and the Aesthetic," in L. Aagaard-Mogensen (ed.) *Culture and Art*, Atlantic Highlands, Humanities Press, 1976, pp. 118–30; Robert McGregor, "Dickie's Institutionalized Aesthetic," *British Journal of Aesthetics*, 1977, vol. 17, pp. 3–13; Michael Mitias, "The Institutional Theory of the Aesthetic Object," *The Personalist*, 1977, vol. 58, pp. 147–55.

17 Walton, op. cit., pp. 350–1. Walton, I suspect, would agree that the psychological claim applies to the aesthetic appreciation of nature. This is not only evidenced by his use of the elephant example, but also by his remark to the effect that what the claim purports "is obviously not an isolated or exceptional phenomenon, but a pervasive characteristic of aesthetic perception" (p. 354).

18 Hepburn, op. cit., p. 55.

19 In discussing the psychological claim, the categories involved are, strictly speaking, the categories of, for example, apparent beaches, apparent seabeds, or apparent elephants, that is, the categories of things which because of their perceptual properties look like beaches, sea-beds, or elephants. See Walton, op. cit., p. 339. Constant repetition of "apparent" is not necessary as long as it is clear that the categories involved are perceptually distinguishable (see footnote 4).

20 Mountains seem a particularly good case in point for testing the psychological claim. See Marjorie Hope Nicolson's classic work *Mountain Gloom and Mountain Glory*, Ithaca, Cornell University Press, 1959.

21 I do not mention Walton's second circumstance, for I think it directly relevant in the case of neither art nor nature. This is because it seems not to be a circumstance

constitutive of correctness as are the other circumstances. In contrast to these, the second circumstance seems only to provide some evidence for correctness. At best it might be construed as a "tie-breaking" consideration somewhat analogous to the way in which simplicity is viewed in regard to theory testing. It is possible that this is also the proper way in which to view the ethical considerations I discuss near the end of this chapter.

22 Walton, op, cit., p. 364, my italics.

23 It may be held that the idea of *discovering*, for example, that a whale is in fact a mammal is simplistic – that the more correct description is in terms of something such as inventing and applying a conceptual system. Although I do not wish to assume any such views here, to the extent that they are plausible, my general line of thought is strengthened. Such views help to bring the categorization of art and the categorization of nature more in line with one another.

24 I owe this example to Donald Crawford. Certain ideas in the present chapter were initially developed as comments on Donald Crawford's unpublished paper "Art and the Aesthetics of Nature," which was presented at the Pacific Division of the American Society for Aesthetics at Asilomar, April 1980. I thank Crawford and other participants for valuable discussion of those ideas.

25 I bring out some of its problematic aspects in "Formal Qualities in the Natural Environment," *Journal of Aesthetic Education*, 1979, vol. 13, pp. 99–114 (reproduced in this volume, Chapter 3).

26 Hepburn, op. cit., p. 62, notes:

> Suppose the outline of our cumulus cloud resembles that of a basket of washing, and we amuse ourselves in dwelling upon this resemblance. Suppose that on another occasion we do not dwell on such freakish aspects, but try instead to realize the inner turbulence of the cloud, the winds sweeping up within and around it, determining its structure and visible form. Should we not be ready to say that this latter experience was less superficial than the other, that it was truer to nature, and for that reason more worth having? If there can be a passage, in art, from easy beauty to difficult and more serious beauty, there can also be such passages in aesthetic contemplation of nature.

Compare Hepburn's general remarks on pp. 60–4 with the argument that follows.

27 Note, for example, the following remark by Aldo Leopold, "Conservation Esthetic" [1953], in *A Sand Country Almanac with Essays on Conservation from Round River*, New York, Random House, 1974, p. 285 (my italics):

> Consider…a trout raised in a hatchery and newly liberated in an over-fished stream…No one would claim this trout has the same value as a wholly wild one caught out of some unmanaged stream in the high Rockies. *Its esthetic connotations are inferior,* even though its capture may require skill.

The inferior aesthetic "connotations" of one trout as opposed to the other can only be a function of what descriptions are true of one as opposed to the other.

28 This illustration was suggested to me by Donald Crawford.

29 I make suggestions concerning this issue in "Appreciation and the Natural Environment," op. cit. (reproduced in this volume, Chapter 4).

6

NATURE AND POSITIVE AESTHETICS

The development of positive aesthetics

In this chapter I examine the view that all the natural world is beautiful. According to this view, the natural environment, insofar as it is untouched by man, has mainly positive aesthetic qualities; it is, for example, graceful, delicate, intense, unified, and orderly, rather than bland, dull, insipid, incoherent, and chaotic. All virgin nature, in short, is essentially aesthetically good. The appropriate or correct aesthetic appreciation of the natural world is basically positive and negative aesthetic judgments have little or no place.

Such a view of the appropriate aesthetic appreciation of the natural world is initially implausible. Concerning the aesthetic appreciation of art, for example, a comparable view would not warrant serious consideration, for it is not the case that all art is essentially aesthetically good; in such appreciation, negative aesthetic judgments have a significant place. In spite of this initial implausibility, the view that all virgin nature is essentially aesthetically good has numerous supporters, particularly among those who have given serious thought to the natural environment. Thus, it deserves careful consideration. In what follows I elaborate it by reference to some historical and contemporary sources and examine various justifications that may be offered for it.

The roots of the view may be traced to at least eighteenth-century ideas concerning the primacy of natural beauty and the beauty of nature as a norm for art.[1] In the nineteenth century it becomes explicit in the writings of landscape artists and others concerned with nature. For example, 1821 marks landscape painter John Constable's much quoted comment: "I never saw an ugly thing in my life."[2] At this time the idea of nature as essentially beautiful is also interwoven with a developing awareness of the negative effects of human intervention. Thus, in 1857 John Ruskin, in giving counsel to landscape artists, can find the certainty of beauty only in that which he takes to be beyond the reach of man:

> Passing then to skies, note that there is this great peculiarity about sky subject, as distinguished from earth subject; – that the clouds, not

> being much liable to man's interference, are always beautifully arranged. You cannot be sure of this in any other features of landscape.[3]

By the second half of the nineteenth century the view can also be found in the work of individuals who are best described as environmental reformers. For example, 1864 dates the publication of George Marsh's *Man and Nature*, which has been described as "the fountainhead of the conservation movement" and "the beginning of land wisdom in this country [the United States]."[4] Here Marsh develops two hypotheses – that "nature left alone is in harmony" and that man is "the great disturber of nature's harmonies." Early in the book, he enumerates nature's "manifold blessings" for man. All, he points out, "must be earned by toil" and are "gradually ennobled by the art of man," with one exception – natural beauty. This blessing is only disturbed, not ennobled, by man; for it, Marsh says, "unaided nature make[s] provision." He suggests that these beauties "that now, even in their degraded state, enchant every eye" could be fully and universally appreciated "only in the infancy of lands where all the earth was fair."[5]

The related ideas of man as the destroyer of nature's beauty and of "all the earth" as beautiful except for man's influence are not limited to somewhat scholarly works such as *Man and Nature*. By the end of the nineteenth century they are key notions in movements of social reform. In public lectures given in the 1880s, artist, poet, and social critic William Morris preaches:

> For surely there is no square mile of earth's inhabitable surface that is not beautiful in its own way, if we men will only abstain from wilfully destroying that beauty; and it is this reasonable share in the beauty of the earth that I claim as the right of every man who will earn it by due labor.[6]

At about the same time, writing in the *Atlantic Monthly*, John Muir promotes wild land preservation with similar themes: "None of Nature's landscapes are ugly so long as they are wild...But the continent's outer beauty is fast passing away."[7]

In the twentieth century these themes, although perhaps neither fully developed nor completely entrenched, have gained a foothold in the public mind. David Lowenthal, a geographer, characterizes the received opinion in our time as follows:

> Nature is...thought preferable to artifice. The favored landscapes are wild; landscapes altered or disturbed or built on by man are considered beneath attention or beyond repair...Conservationist organizations contrast sordid scenes dominated by man with lovely landscapes devoid of human activity...The implication is clear: man is dreadful, nature is sublime.[8]

The most recent expressions of the view that the natural world is essentially aesthetically good are found in philosophical writings that address environmental issues. Unlike artists and reformers, however, contemporary philosophers are not given to absolute or universal claims. Nonetheless, there is what Leonard Fels describes as one "basic assumption for aesthetic judgments." It is "that anything natural is probably good. Much that is natural is also thought of as inherently beautiful."[9] This attribution of aesthetic goodness, inherent beauty, or more particular positive aesthetic qualities to all that is natural is illustrated by a number of current discussions. For example, in considering the preservation of species, Lilly-Marlene Russow, while expressing worries about "species such as the snail darter," claims that:

> ...aesthetic value can cover a surprising range of things: a tiger may be simply beautiful; a blue whale is awe-inspiring; a bird might be decorative; an Appaloosa is of interest because of its historical significance; and even a drab little plant may inspire admiration for the marvelous way it has been adapted to a special environment.[10]

The same idea occurs in attempts to analyze the qualities and values that justify preservation. For example, Kenneth Simonsen holds:

> There is...something astonishing in this world which has been brought into being by obscure if not blind forces...It is perhaps this realization which is at the root of our wonder at wild things. Once this attitude comes to pervade our response to nature, we feel more than admiring respect. Rather, for us, all wild things become invested with a sense of awe.[11]

Similarly, Holmes Rolston argues:

> Wild nature has a kind of integrity...The Matterhorn leaves us in awe, but so does the fall foliage on any New England hillside, or the rhododendron on Roan Mountain. Those who linger with nature find this integrity where it is not at first suspected, in the copperhead and the alligator, in the tarantula and the morel, in the wind-stunted banner spruce and the straggly box elder, in the stormy sea and the wintry tundra...This value is often artistic or aesthetic, and is invariably so if we examine a natural entity at the proper level of observation or in terms of its ecological setting. An ordinary rock in micro section is an extraordinary crystal mosaic. The humus from a rotting log supports an exquisite hemlock...Natural value is further resident in the vitality of things, in their struggle and zest, and it is in this sense that we often speak of a reverence for life, lovely or not. Or should we say that we find all life beautiful.[12]

Other contemporary writers defend a related but more moderate point of view, arguing that although not every kind of thing in the natural world is aesthetically good, being natural is nonetheless essentially connected with positive aesthetic qualities and value. For example, Joseph Meeker, while suggesting that a "burned forest is ugly because it represents a truncated system of growth,"[13] in general holds that:

> The human experience of beauty is rooted in natural forms and processes...and our esthetic values are really no more – nor any less – than abstract formulations of the natural...What is exclusively human is generally not beautiful to us, and we find beauty in human art only when it is compatible with forms and processes in nature.[14]

The moderate view is developed in more detail by Robert Elliot. Elliot explicitly states that "I do not want to be taken as claiming that what is natural is good and what is non-natural is not. The distinction between natural and non-natural connects with valuation in a much more subtle way than that."[15] He elaborates: "Sickness and disease are natural in a straightforward sense and are certainly not good. Natural phenomena such as fires, hurricanes, volcanic eruptions can totally alter landscapes and alter them for the worse."[16] With this qualification, Elliot's view is that "the naturalness of a landscape is a reason for preserving it, a determinant of its value...What the environmentalist insists on is that naturalness is one factor in determining the value of pieces of the environment."[17] He argues for this view by reference to analogies between artistic fakes and "faked" nature.[18]

Reflection on these and related themes in the aesthetic appreciation of nature has led Aarne Kinnunen to sharply distinguish the aesthetics of nature from the aesthetics of art. He claims the former is positive and the latter critical. Critical aesthetics, he argues, allows for negative aesthetic criticism and is appropriate for art, and for nature only when it has been affected by man. Positive aesthetics, on the other hand, does not involve negative aesthetic judgment, but only the acceptance and aesthetic appreciation of something for what it is. According to Kinnunen, the aesthetics of nature is essentially positive in that all virgin nature is beautiful. He writes:

> ...all untouched parts of nature are beautiful. To be able to enjoy nature aesthetically is distinct from judgment. The aesthetics of nature is positive. Negative criticism comes into play only when man's part in affecting nature is considered.[19]

Kinnunen's remarks summarize the pattern of thought I have elaborated here. Consequently I label the view under consideration in this article the "positive aesthetics" position.

Nature appreciation as non-aesthetic

In this and the following two sections I consider three possible justifications for positive aesthetics. The first is the nonaesthetic view of nature appreciation introduced in Chapter 1. This view holds that our appropriate appreciation of the natural world is essentially as the positive aesthetics position indicates, except that it is not genuinely aesthetic. Consequently, the initial implausibility of the position vanishes: if the appreciation in question is not aesthetic, it need not involve the critical, negative judgments so common in the aesthetic appreciation of art. Indeed, the perceived lack of this judgmental element is taken as the reason for denying that our appreciation of nature is aesthetic.

Of the individuals discussed in the previous section, Elliot most clearly represents this point of view. He holds that "certain concepts which are frequently deployed in aesthetic evaluation" can "usefully and legitimately be deployed in evaluations of the environment." He says, for example, that we might admire "the intricate and delicate shadings in a eucalyptus forest" or "may be awed by a mountain." Yet he claims that such "responses to nature" do not "count as aesthetic responses." His reason for this lies in the aspect of appropriate aesthetic appreciation emphasized by positive aesthetics. He notes an argument that "turns on the claim that aesthetic evaluation has, as a central component, a judgmental factor" and contends that according to this argument "the judgmental element in aesthetic evaluation serves to differentiate it from environmental evaluation." Of the judgmental factor, he says:

> ...an apparently integral part of aesthetic evaluation depends on viewing the aesthetic object as an intentional object, as an artifact, as something that is shaped by the purposes and designs of its author. Evaluating works of art involves explaining them, and judging them, in terms of their author's intentions; it involves placing them within the author's corpus of work; it involves locating them in some tradition and in some special milieu. Nature is not a work of art.[20]

The basic line of argument may be reconstructed as follows: first, since the natural world does not have the features sketched in this quote, environmental evaluation cannot have the "apparently integral part of aesthetic evaluation" which depends upon them. Thus, second, since environmental evaluation lacks this judgmental element, it is not aesthetic evaluation. And, third, this in turn is grounds for concluding that our responses to nature are not aesthetic responses. This conclusion is not unique to Elliot. Its more general acceptance is suggested by Harold Osborne's observation that some contemporary aestheticians "doubt whether we are talking about the same thing when we speak of natural beauty and the beauty of fine art."[21]

I think Elliot's conclusion is implausible. It stands in conflict with the commonly held view that everything is open to aesthetic appreciation (although not necessarily positive aesthetic appreciation).[22] If this common view is correct, it would at least be odd if our appreciation of nature were never aesthetic appreciation and our responses never aesthetic responses. This can be brought out by noting certain ramifications of the view. For example, Paul Ziff not only includes non-artifacts, but every kind of non-artifact within the scope of aesthetic appreciation. He uses unparadigmatic examples as "a gator basking in the sun on a mud bank in a swamp" and "a mound of dried dung."[23] In a similar vein, environmentalist Neil Evernden assumes that an "individual may achieve aesthetic involvement with a spruce bog, a prairie, or a charred forest. The potential for this phenomenon is inherent in every landscape."[24] Such diverse examples suggest the equality of all things as proper objects of aesthetic appreciation. Ziff makes this explicit by a comparison between the appreciation of his basking gator and that of Leonardo Da Vinci's *Ginevra de Benci* (1504), concluding that "Anything that can be viewed can fill the bill of an object fit for aesthetic attention and none does it better than any other."[25] Moreover, certain more typical instances of the appreciation of nature – for example, the appreciation of colorful sunsets or delicate wild flowers – are usually thought to be paradigmatic instances of aesthetic appreciation, instances in terms of which we acquire and understand the concept of the aesthetic. To deny that these are cases of aesthetic appreciation comes close to challenging the coherence of that concept.

Having briefly considered the conclusion, let us turn to the argument itself. The argument is problematic at each of its three steps. First, in step three, even if environmental evaluations are not aesthetic, it does not follow that responses to nature are not aesthetic responses. Genuine aesthetic responses to nature might occur independently of any evaluations of nature, or they might support nonaesthetic environmental evaluations in something like the way in which genuine taste responses (for example, tasting the sweetness of milk) can support non-taste evaluations of what is tasted (for example, this milk is good to drink). Step three of the argument seems to involve a non sequitur.

Second, in step two, even if environmental evaluations lack the "judgmental element," it does not follow that they are not aesthetic – at least not without more argumentation. On the one hand, if the "judgmental element" refers to the critical, negative judgments referred to in the positive aesthetics position, then it is possible that positive aesthetics is simply correct – that all virgin nature is simply aesthetically good. In this case the judgmental element simply does not arise in our appropriate appreciation of nature. If we created a genre of art which gave rise to no negative judgments, it would not follow that our appreciation and evaluations of it were therefore not aesthetic. On the other hand, if, as Elliot seems to hold, the "judgmental element" refers just to the particular kind of critical judgments we can make only in regard to art, it still does not follow that environmental evaluations are not aesthetic; it only follows that they are not evaluations based on the kinds of judgments we can make only with regard to art. But why should they be based on these kinds of judgments? They are not evaluations of art. As Elliot reminds us: "Nature is not a work of art."

Third, in step one, even if the natural world is not an intentional object, not an artifact, not something that is shaped by the purposes and designs of its author, it does not follow that environmental evaluation cannot have the "apparently integral part of aesthetic evaluation" which depends upon such features.[26] This is because although this "part" may depend upon these features in the case in question – aesthetic evaluations of art – it need not depend upon them in other cases. If the object of aesthetic appreciation is an intentional object, an artifact, something that is shaped by the purposes and designs of its author, then aesthetically evaluating it may indeed involve explaining it and judging it in terms of the author's intentions, placing it within the author's corpus of work, locating it in some tradition and in some special milieu. But if it is not, then aesthetically evaluating it need only involve something analogous to this. If it is a wildflower or a basking gator, aesthetically evaluating it may involve explaining it and judging it in terms of what it is, placing it within its natural category, its species, genus, etc., locating it in its natural history and in its environmental milieu.

I return to these kinds of considerations in a later section. At present I end this examination of the argument with another of Ziff's remarks:

> That something is not an artifact does not suggest let alone establish that it is therefore unfit to be an object of aesthetic attention. And unless one has a compelling narcissistic obsession with the marks of men's endeavors one can view things in the world aesthetically without being concerned with or inhibited by their lack of status as artifacts.[27]

I conclude that attempting to account for positive appreciation of the natural world by denying that this appreciation is aesthetic does not succeed. The claim that our appreciation of nature is not aesthetic is implausible and at least one line of argument for that claim is inadequate.

Positive aesthetics and sublimity

A second defense of positive aesthetics takes the fact that the natural world is not an artifact – that it does not have an artist – not as grounds for claiming that our appreciation of it is not aesthetic, but rather as a justification for the view that it is appropriate to appreciate it aesthetically only in a positive manner. The idea is that the natural world, by not having an artist, is in an important sense outside the bounds of human control. This is meant to account for the particular aesthetic appreciation we have of virgin nature and for the claim that negative aesthetic criticism of it is out of place.

The defense is conceptually and perhaps historically related to the tradition of appreciating wild nature in terms of the sublime, especially as developed in the theories of Burke and Kant.[28] The appreciation of nature as sublime is a function of the fact that since it is outside the bounds of human control, it is potentially a threat, a source of fear or terror. However, as an element of aesthetic experience the threat dissolves into positive aesthetic appreciation: amazement, wonder, or awe in the face of nature's threatening otherness. In contemporary thought the idea is applied not simply to wild nature because of its threatening otherness, but to all the natural world because of its astonishing qualities.

Of the individuals mentioned in the second section of this chapter, Simonsen most clearly represents this line of thought. After discussing Burke, he concludes that "Burke has confounded the astonishing with the terrifying."[29] He then describes the ideal nature appreciator (who happens to be an atheist):

> He is confronted with a natural world which has come into existence on its own, and not in accordance with the design of an intelligent creature. He cannot enter into this world, as he can the world of human fabrication. There is, therefore, something astonishing in this world which has been brought into being by obscure if not blind forces. The wild spectacle he sees before him is truly full of wonder; it just appears, and stands before him like some terrible goddess...an Aphrodite emerging mysteriously from formless seas. It is perhaps this realization which is at the root of our wonder at wild things. Once this attitude comes to pervade our response to nature...all wild things become invested with a sense of awe.[30]

In this view the positive aesthetic appeal of the natural world is a function of the astonishing nature of something that is beyond the limits of human control. Moreover, this is also the basis for the claim that negative criticism has no role. This aspect of the view can be brought out by noting that virgin nature is beyond human control in two senses. First, it is beyond human beings' physical control in that they do not bring it into being and, insofar as it remains virgin nature, they do not maintain, shape, design, or develop it. Second, moreover, the natural world is also in a sense beyond human beings' mental control. We understand little of why it exists, how it maintains itself, and what it is really like. Its existence is not only "not in accordance with the design of an intelligent creature," it is also "brought into being by obscure if not blind forces." It "just appears" like "an Aphrodite emerging mysteriously from formless seas." We "cannot enter into this world" as we "can the world of human fabrication." It is alien in the sense that we do not bring it into being, and it is obscure in the sense that we do not really understand it.

These two senses in which virgin nature is beyond human control together support the view that negative aesthetic criticism of it is inappropriate, in that it is both pointless and presumptive. It is pointless because it will not, in fact logically cannot, change virgin nature. There is no artist, no intelligence, who might in light of such criticism redesign and alter the existing natural world or even change the design and create a new one, thereby improving our aesthetic situation. Moreover, such criticism is presumptive in light of our limited understanding. We have no grounds for negative criticism of something that is as alien and obscure as is virgin nature, that is not even in accordance with the design of any intelligence. In Simonsen's remarks there is even the suggestion that this would be disrespectful. What right do we have to criticize some mysterious and "terrible goddess"? The only appropriate response is wonder and awe.

This view partially supports the positive aesthetics position, but on balance it is not an adequate justification. It does aid in understanding certain positive aesthetic qualities of nature. However, if virgin nature is not to be appreciated aesthetically in a negative manner, this is not because doing so is either pointless or presumptive. This can be brought out by analogy with the aesthetic appreciation of art. First, consider works of art that are by a long dead artist and belong to a genre within which artists have ceased to work; perhaps the techniques for working within it have been lost in time. Such works, like any other, would be open to and subjected to negative aesthetic criticism. However, there is no artist who might in light of such criticism redesign and alter the existing works or even change the design and create new works in this genre. If some existing artist did alter the works they would no longer be the same works, and if he or she created new works they would not be works within this genre. Negative criticism is here as pointless as it is claimed to be concerning virgin nature, but this does not alter the nature of art appreciation thereby making it positive aesthetics. It may be objected that the art case is nonetheless distinct because the original works could have been different, could have been aesthetically better than they are, and this fact gives a point to criticism. However, such an objection is not persuasive because, for all we know, the natural world also could have been different, could have been aesthetically better than it is. In fact, that it could have been seems very likely. If this gives a point to criticism in the art case, it does the same for criticism of the natural world.

Second, concerning presumptiveness, imagine that the above-mentioned works are also hopelessly obscure and belong to a hopelessly obscure genre. Such art would still be open to and subjected to negative aesthetic criticism. In fact, some of such criticism would be a result of their obscurity. However, this criticism need not be viewed as presumptive, but rather the opposite, for some of it would function as a means of coming to grips with the works – a means of reducing their obscurity. This may reflect the way in which we develop an appreciation for much rather obscure modern art. It may be objected here that the art case is yet distinct because the works of art, since they are art and the products of human intelligence, are not alien and therefore not essentially obscure as is the natural world. Consequently, it is possible to gain an understanding and appreciation of them and for this reason negative aesthetic criticism is not presumptive. I agree that in some sense the natural world and not art is alien, but do not agree that the natural world is therefore any more essentially obscure than art is. The crucial error is concluding that because the natural world is alien in the sense of not being of our making, it is therefore beyond our understanding. Although the natural world, it may be granted, is not an artifact, not "in accordance with the design of an intelligent creature," this does not mean we cannot understand it. It only means that we cannot come to understand it as we can an artifact, that is, in virtue of creating it. Rather we come to understand the natural world in different ways. (I return to this point in the last section of this chapter.)

I conclude that the appeal to pointlessness and presumptiveness does not establish that negative aesthetic criticism of the natural world is inappropriate. Thus, I argue, neither the tradition of the sublime nor its contemporary sequels provide the means for justifying positive aesthetics.

Positive aesthetics and theism

Each of the two justifications examined above account for positive aesthetics by reference to the fact that the natural world is not a human artifact. A third justification also holds that nature is not of human making. It relies on the theist view that the natural world is designed, created, and maintained by an all-knowing and all-powerful God. This view is familiar enough that it need not be elaborated here. I concentrate on it only as a possible justification for positive aesthetics.

The justification can be developed in two ways. The first takes the divine origin of nature as grounds for positive aesthetics by reference to pointlessness and presumptiveness, although perhaps presumptiveness plays the greater role. It is not necessary here to consider this development of the "divine justification" for two reasons: first, it appears either to have little contemporary support or, when it does, to have a structure similar to the justification considered in the previous section. Second, in light of this, the arguments advanced in that section, or analogous arguments, should be adequate to dismiss it from serious attention.

The second way utilizes the idea that the natural world, since it is of divine design and origin, is a perfect world, at least insofar as it has not been marred by humans. Virgin nature is, therefore, only aesthetically good and closed to negative aesthetic criticism. In discussing the differences between art appreciation and nature appreciation, Nelson Potter characterizes a part of this view:

> The theist sees the world as throughout the product of God's design and plan. For example, he may regard a spectacular sunset as having been rigged by God for the human observer's pleasure and appreciation, just as the same observer views paintings in an art gallery as having been composed for his pleasure and appreciation...The only difference is that in nature, the Artist is divine, not human.[31]

Furthermore, Simonsen points out that if one accepts this kind of view, "one could certainly see why...we would have adequate cause to delight in its [nature's] existence, for it would be an objectification of the divine mind."[32] There is no doubt that if one accepts the basic theist view, it appears to provide a justification for the positive aesthetics position. However, aside from the issue of its acceptability, the theist view as a justification for positive aesthetics presents certain puzzles.

First, it seems that if the divine origin of the natural world is the justification of the positive aesthetics position, then only theists would either justifiably believe in its truth or appreciate the natural world appropriately. This conclusion seems doubtful, but it could possibly be true. However, the divine justification further suggests that there is a very significant difference between the aesthetic appreciation of the natural world of the theist and that of the nontheist. Potter elaborates certain ramifications of this. He holds that the view that natural beauty is a "product of God's design and plan…seems to require a radically different kind of appreciation of such beauty for the atheist."[33] He points out:

> And yet the theist who hears that an acquaintance is an atheist, would surely not thereafter assume that the acquaintance has a radically different kind of appreciation of the beauties of nature (or at least many theists would not). Or if the theist hears from a friend that the friend shares the theist's intense enjoyment of the beauties of nature, could the theist rightly infer that his friend must also be a theist…? Surely not…In general, any kind of inference from the aesthetic enjoyment of nature to one's religious views, or any inference in the other direction must be quite dubious.[34]

Potter concludes that because the view that natural beauty is a product of God's design "entails that many such inferences are not dubious, we have reason for thinking that that view is incorrect."[35] I am not certain that Potter's observations are a reason for thinking the view incorrect, but they do make clear some of the oddity of the divine justification.

A second puzzling feature of this justification can be seen in relation to a traditional problem in Western thought: the problem of evil. The problem is how to reconcile an all-knowing, all-powerful, and all-moral deity with the existence of evil in the world that he or she has created. The evil is often divided into moral evil, which is of human making, and natural evil, which is not. The latter is typically illustrated by natural catastrophes; it is all that need concern us here. Attempts to solve the problem of evil fall into three categories: first, denials of the existence of evil, second, denials of one or other of the deity's three problematic properties and, third, attempts at theodicy, that is, explanation of why a deity with these three properties would nonetheless allow there to be evil in the world. Concerning natural evil, theodicy frequently argues that evil is necessary to human beings for various purposes, such as the recognition of good, the development of virtue, and so forth.

It is easy to posit an analogous problem of ugliness. Since an all-knowing, all-powerful, and all-moral deity would presumably have perfect aesthetic judgment, how is he or she to be reconciled with the existence of ugliness or of more specific negative aesthetic qualities in the world that he or she has created? The analogous problem is amenable to conceptually parallel solutions: first, deny the existence of ugliness, second, deny one or other of the deity's three (or four) problematic properties and, third, engage in "aesthetic theodicy." If, in the problem of ugliness, as in the problem of evil, we distinguish ugliness of human making and natural ugliness, it becomes clear that positive aesthetics is a position of the kind embraced by the first type of solution. Or to put it another way, the justification of the positive aesthetics position by reference to the divine origin of the natural world constitutes a pattern of thought equivalent to solving the problem of (natural) ugliness simply by denying the existence of such ugliness. This begins to reveal a further oddity of this justification. On the one hand, it seems straightforwardly circular, and, on the other, counter-intuitive. In the problem of evil, the parallel solution, the denial of the existence of evil, and especially of natural evil, is considered the least acceptable and most implausible line of thought. Theists in the Western tradition have almost without exception pursued theodicy, although the occasional philosophical theist has flirted with solutions of the second type.[36] This suggests that the theist should engage in aesthetic theodicy rather than deny the existence of natural ugliness. In short, the theist, who accepts the justification's premise that the natural world is of divine origin, seems an unlikely candidate for accepting its conclusion that the natural world is essentially beautiful.

A third set of considerations relevant to the divine justification concerns the historical relationship between Western theism and the positive aesthetics position. As a matter of history, theists have not usually accepted this position and theism, at least as embodied in Christianity, has given little or no support to it. This, of course, is a complex and subtle historical claim about which I only make a few relevant observations. First, many authorities believe that Christian theism traditionally viewed wild nature as something to be confronted, dominated, and domesticated by human beings for their purposes.[37] Given this, it is understandable that the theist would view tamed and cultivated nature rather than virgin nature with aesthetic favor. Within such a climate of thought it is difficult to see how positive aesthetics would develop, even if it could be justified by the natural world's assumed divine origins. Moreover, it has been noted that Christian thought developed such as to give support to the opposite point of view – the view that virgin and wild nature is aesthetically repellent. For example, in discussing the aesthetic appreciation of mountains, Ronald Rees, a geographer, points out that until the end of the seventeenth century:

> Many theologians regarded the world, as they did man himself, as fallen from a state of pristine innocence. It was commonly believed that the original world had been a perfectly smooth sphere – the

> 'Mundane Egg' – whose surface, as punishment for man's sins, had been ruptured by interior fluids piling up, to quote one seventeenth-century theologian, 'vast and undigested heaps of stones and earth', those 'great ruins' we call mountains.[38]

Rees is quoting from Bishop Thomas Burnet's *Sacred Theory of the Earth* which was written to give, in Burnet's words, "some tolerable account of how confusion came in nature."[39] It is informative to view Burnet as attempting to deal with the problem of natural ugliness by means of aesthetic theodicy. Historically it seems that this line of thought rather than positive aesthetics was cultivated by Christian theism.

Another historical consideration is the contention that in Western culture the aesthetic appreciation of nature has developed in inverse proportion to religion. For example, Russian historian Victor Romanenko claims, concerning landscape appreciation, that:

> Strictly speaking, neither Antiquity, nor the Middle Ages, nor the early Renaissance had any idea of landscape. Landscape appeared at a time when criticism of the religious world outlook had ripened in the minds of men, who more acutely than ever before became aware of their unity with nature.[40]

Concerning aesthetic appreciation in general, Romanenko notes a world view that "inevitably limits" man's "aesthetic attitude towards nature" and "retards the development of his feelings for nature." He adds:

> Christianity, as it were, sustains and conserves such a world view, because religion plays down nature, reduces her to the position of a most christian subject of the all-mighty creator. But this only so long as science is in its infancy and still under the guardianship of religion.[41]

The claim is that Christianity and the aesthetic appreciation of nature have been opposing forces to an extent such that the latter could grow only as the former went into decline. If this is an accurate historical picture, it is not surprising to find that positive aesthetics does not prosper under Christian theism.

I conclude that the divine justification is not promising. It not only suggests that the theist has a unique kind of aesthetic appreciation of nature, but is counterintuitive in light of both the theist position on the problem of evil and the historical point of view of Christian theism. Perhaps to find a justification for positive aesthetics we need to look elsewhere, not to religion but rather, as suggested by Romanenko's quote, to the infant long under religion's guardianship: science.

Science and aesthetic appreciation of nature

In the remaining sections of this chapter I explore the role of natural science in our aesthetic appreciation of nature. Consideration of science suggests a more plausible justification for positive aesthetics than any of those examined in the last three sections.

The importance of science in the development of the aesthetic appreciation of nature is brought out in Marjorie Hope Nicolson's classic *Mountain Gloom and Mountain Glory*. Arguing that appeal to the history of art does not provide the complete account, Nicolson elaborates the growth of the "aesthetics of the infinite."[42] Here science plays the essential role. Seventeenth-century developments in astronomy and physics, followed by those in geology and geography, explain and expand the natural world such that the notion of the sublime can find a place in its appreciation. Wonder and awe, formerly thought appropriate only for a deity, now become aesthetic responses to the seemingly infinite natural world; and especially to landscapes such as Nicolson's prime example, mountainscapes. Moreover, all landscapes become objects of appreciation as a function of the continued advances of the natural sciences. Speaking of landscape painting as a reflection of landscape appreciation, Romanenko claims: "Realistic landscape…reached a flourishing stage in the nineteenth century and…its development was definitely related to the rapid advance in all fields of natural science, especially geography, biology, geology."[43]

In the nineteenth century, in addition to geography and geology, biology becomes particularly important. Scientists of that century, especially Darwin, turned the previous practice of collecting and naming into the science of the day, thereby accomplishing for flora and fauna what sciences such as astronomy and geology had earlier accomplished for landscapes:

> Everything, which before the days of Darwin had borne the stamp of "divine origin", the beauty of nature included, was passed down to the earth from heaven. The idea of man's unity with nature, since the days of Darwin, received particularly wide recognition, and his evolutionary theory, his methodology and the principles of his historical approach began, in the latter part of the nineteenth century, to exercise a revolutionizing influence on a subject so remote from exact natural science as aesthetics.[44]

With the growth of our aesthetic appreciation of nature in general science shares credit with art. However, in the more recent development of positive aesthetics, science is the main if not exclusive factor. The individuals who hold the positive aesthetics position take science to have a special relevance to their views. This is particularly true of Marsh; simply a glance at *Man and Nature* reveals the importance in his thought of both the theoretical and the applied science of his day. Moreover, Constable and Ruskin can be seen as no less under the spell of science. Constable is famous for his claim that "painting is a science, and should be pursued as an inquiry into the laws of nature." He asks: "Why, then, may not landscape painting be considered as a branch of natural philosophy, of which pictures are but the experiments?"[45] Equally significant is the following remark: "I must say that the sister arts have less hold on my mind…than the sciences, especially the study of geology, which, more than any other, seems to satisfy my mind."[46] And it is well known that Ruskin's descriptions of nature compare favorably with those of his scientific contemporaries in accuracy and detail. Ronald Rees claims that "parts 2, 5 and 7 of 'Modern Painters' in effect offer a course in physical geography for landscape painters."[47]

Contemporary advocates of positive aesthetics have a similar regard for science. The main difference is that now the relevant sciences are not primarily geography and geology, but rather biology and the all encompassing "science" of ecology. The issues these writers frequently address, such as the survival of species and the value and preservation of wild lands, are themselves issues of biological and ecological importance. Their orientation is even revealed in their titles: "Aesthetic Decision-Making and Human Ecology," "The Biology of Beauty," "Can and Ought We to Follow Nature?" "Is There an Ecological Ethic?" "The Value of Wildness," "Ecological Esthetics."[48] More significant is the way they bring biological and ecological insight to the positions they develop. For example, in discussing "languages" of art and "languages" of nature in aesthetic appreciation, Kinnunen characterizes the latter as "ecological languages."[49] Elliot endorses the view that "the understanding of the complexity, diversity, and integration of the natural world which ecology affords us, opens up a new area of valuation."[50] Rolston in speaking of "wild value" claims that "this value is often artistic or aesthetic," but "also has to do with the intelligibility of each of the natural members; and here natural science, especially ecology, has greatly helped us."[51] He also argues that "environmental science" invites us "to see the ecosystem not merely in awe, but in 'love, respect, and admiration.' "[52] As Meeker suggests: "Esthetic theory may be more successful in defining beauty when it has incorporated some of the conceptions of nature and its processes which have been formulated by contemporary biologists and ecologists."[53] It seems no exaggeration to say that this is what these individuals are attempting to do.

That science has played a role in the development of both the aesthetic appreciation of nature and the positive aesthetics position is, I think, clear. What is less clear is exactly why this should be so, although in general the answer is obvious enough: science provides knowledge about nature. For example, even though Biese's *The Development of the Feeling for Nature in the Middle Ages and Modern Times* is mainly an account of art's role in nature appreciation, Biese concludes that the "present intensity" (1905) of the feeling for nature "is due to the growth of science, for although feeling has become more realistic and matter-of-fact in these days of electricity and the microscope, love for Nature has increased with knowledge."[54] In a similar vein Val Routley argues that the knowledge that science has gathered within the last three hundred years and that has only recently been generally disseminated "makes natural areas intelligible" and "as usual, information is an important adjunct to appreciation, as much so with the natural area as with the string quartet."[55] Following Routley, Elliot takes note of the knowledge provided by ecology. Of those who value wilderness, he says:

> What they do see, and what they value, is very much a function of the degree to which they understand the ecological mechanisms which maintain the landscape and which determine that it appears the way it does. Similarly, knowledge of art history, of painting techniques, and the like will inform aesthetic evaluations and alter aesthetic perceptions. Knowledge of this kind is capable of transforming a hitherto uninteresting landscape into one that is compelling.[56]

I think there is no doubt that scientific knowledge, as Elliot suggests, can transform the landscape, in fact the natural world, but there is yet the question of how and why this transformation occurs: how and why does scientific knowledge make the natural world seem beautiful? The beginnings of an answer are suggested by Rolston when he says:

> Ecological description finds unity, harmony, interdependence, stability, etc.…earlier data are not denied, only redescribed or set in a larger ecological context, and somewhere enroute our notions of harmony, stability, etc., have shifted too and we see beauty now where we could not see it before.[57]

Similarly Routley suggests: "The informed person…sees a pattern and harmony where the less informed may see a meaningless jumble."[58] Scientific information and redescription make us see beauty where we could not see it before, pattern and harmony instead of meaningless jumble. If these suggestions are correct, they begin to explain the relationship between scientific knowledge and the aesthetic appreciation of nature. They begin to account for the way in which the two have developed hand in hand and why, in light of scientific knowledge, the natural world seems aesthetically good. However, as they stand they do not do justice to the complexity of aesthetic appreciation, and by themselves they do not provide a justification for the positive aesthetics position.

Science and appropriate aesthetic appreciation of nature

In this section I reiterate and elaborate those observations about appropriate aesthetic appreciation and appropriate appreciation of nature that are introduced and defended in the preceding two chapters.

Appropriate aesthetic appreciation is that appreciation of an object which reveals what aesthetic qualities and value it has. In the aesthetic appreciation of art, it may be assumed that relevant knowledge is essential to such appreciation. As Kendall Walton says:

> If we are confronted by a work about whose origins we know absolutely nothing (for example, one lifted from the dust at an as yet unexcavated archaeological site on Mars), we would simply not be in a position to judge it aesthetically. We could not possibly tell by staring at it, no matter how intently and intelligently, whether it is coherent, or serene, or dynamic, for by staring we cannot tell whether it is to be seen as a sculpture, a guernica [a category of *Guernica*-like works invented by Walton], or some other exotic or mundane kind of work of art.[59]

The suggestion here is that, as elaborated in Chapter 4, in order to appreciate what aesthetic qualities a work has, it is necessary to know how it is to be seen. And knowing how it is to be seen is a function of knowing what it is and something about it. For example, Van Gogh's *The Starry Night* (1889) is a dynamic and vibrant post-impressionist painting. However, were it seen as a German expressionist work, it would appear more serene, somewhat subdued, even a bit dull. In order to appropriately appreciate it, to appreciate its dynamic and vibrant qualities, we must see it as a post-impressionist painting. This requires knowing that it is a post-impressionist painting and something about post-impressionist painting; it requires the kinds of knowledge given by art history and criticism.

These observations are developed by different individuals in different ways. As noted in Chapter 4, Ziff captures the idea of appropriate aesthetic appreciation by distinguishing different "acts of aspection." He contends: "Generally speaking, a different act of aspection is performed in connection with works belonging to different schools of art, which is why the classification of style is of the essence."[60] As spelled out in detail in Chapter 5, the importance of classification is further developed by Walton who introduces the notion of "categories of art." "Such categories include media, genre, styles, forms, and so forth – for example, the categories of paintings, cubist paintings, Gothic architecture, classical sonatas."[61] Walton points out that works of art can be "perceived in" different categories and only by perceiving them in the correct categories are we likely to achieve an appropriate aesthetic appreciation of them. Correctness of categories, Walton further argues, is in part determined by certain facts about the origins of works, in particular, in what category their artists intend them to be perceived and what categories are well established and recognized by the societies in which they are produced.

In the aesthetic appreciation of art, I take the kind of view indicated by Ziff and Walton to be essentially correct. For present purposes I assume a position similar to Walton's and utilize some of his terminology. What needs to be shown here is that an analogous account applies to the aesthetic appreciation of nature. The analogous account holds that there are different ways to perceive natural objects and landscapes. This is to claim that they, like works of art, can be perceived in different categories – not, of course, in different categories of art, but rather in different "categories of nature." Analogous to the way *The Starry Night* might be perceived either as a post-impressionist or as an expressionist painting, a whale might be perceived either as a fish or as a mammal, or a gator basking in the sun might be perceived either as an alligator or as a crocodile. Further, for natural objects or landscapes some categories are correct and others not. As it is correct to perceive the Van Gogh as a post-impressionist painting, it is likewise correct to perceive the whale as a mammal and the gator as an alligator. Lastly, analogous to the way certain facts about works and their origins in part determine the correct categories of art for them, certain facts about natural objects or landscapes and their origins in part determine the correct categories of nature for them. As certain facts about the Van Gogh and its history in part determine it to be a post-impressionist painting, so certain facts about the whale and its natural history in part determine it to be a mammal. Concerning the gator Ziff says the following:

> Consider a gator basking in the sun on a mud bank in a swamp. Is he a fit object for aesthetic attention? He is and that he is is readily confirmable. Go look and see if you doubt what I say. He is presently to be seen around Chokoloskee Island in the Everglades. What is in question is the American alligator (*Alligator mississipiensis*) not to be confused with a crocodile. Gators have shorter broader heads and more obtuse snouts. The fourth enlarged tooth of a gator's lower jaw fits into a pit formed for it in the upper jaw whereas a crocodile's fits into an external notch. It helps in viewing a gator to see it as a gator and not as a crocodile. But that requires knowing something about gators.[62]

The account of the appropriate aesthetic appreciation of nature suggested here can be elaborated as follows: first, as in the case of art, the aesthetic qualities natural objects and landscapes appear to have depend upon how they are perceived. The rorqual whale is a graceful and majestic mammal. However, were it to be perceived as a fish, it would appear more lumbering, somewhat oafish, perhaps even a bit clumsy (maybe somewhat like the basking shark). Similarly the graceful and even elegant moose would seem an awkward deer; the charming, cute woodchuck, a massive and awe-inspiring brown rat; the delicate sunflower, a stiff and stodgy daisy. Concerning landscapes, consider the example from Ronald Hepburn quoted in detail in Chapter 5 in which he notes that, for example, a "wide expanse of sand and mud" may appear to have different aesthetic qualities – "a wild, glad emptiness" as opposed to "a disturbing weirdness" – depending upon whether it is perceived as a beach or as a tidal basin.[63] Second, as in art appreciation, in these cases, to appropriately appreciate the objects or landscapes in question aesthetically – to appreciate their grace, majesty, elegance, charm, cuteness, delicacy, or "disturbing weirdness" – it is necessary to perceive them in their correct categories. This requires knowing what they are and knowing something about them – in the cases in question, something of biology and geology. In general, it requires the knowledge given by the natural sciences.

I think the above is essentially the correct account of the appropriate aesthetic appreciation of nature. As noted, it is the account that I defended in more detail in the preceding two chapters.[64] Moreover, it has plausibility simply in virtue of being analogous to the assumed position in the aesthetic appreciation of art. If we do not have an account that is analogous in this sense, we may be forced to accept one of two alternatives: either that while the appreciation of art is aesthetic, the seemingly similar appreciation of nature is not; or that even though they are both aesthetic, they are yet different in their natures and/ or structures. Each of these alternatives is implausible. I have argued for the implausibility of the first alternative in third section of this chapter and have brought out some of the problems inherent in the second alternative in the fourth and fifth sections. In the account sketched here, however, our appreciation of nature is aesthetic and is analogous to that of art in both its nature and its structure. The significant difference is that while, in art appreciation, categories of art and the knowledge given by art criticism and art history are relevant, in nature appreciation, the categories are natural categories and the knowledge is that provided by natural history – by science. But this difference is not unexpected; nature is not art.

In the immediately preceding section, I indicated how the growth of natural science has promoted the aesthetic appreciation of nature and how scientific knowledge is especially effective in enhancing it. According to the account outlined here, scientific knowledge is essential for appropriate aesthetic appreciation of nature; without it we do not know how to appreciate it appropriately and are likely to miss its aesthetic qualities and value. Thus, if this account is correct, it explains the ability of science to both promote and enhance aesthetic appreciation of the natural world. Within the context of this chapter, the account also has another virtue: it suggests a means by which to justify positive aesthetics.

Science and positive aesthetics

Positive aesthetics claims that the natural world is essentially aesthetically good. It follows that the natural world must appear as such when it is appropriately aesthetically appreciated. If the view sketched above is to account for this, the natural world must appear aesthetically good when it is perceived in its correct categories, those given and informed by natural science. If this is the case and we can understand how and why it is, we thereby have a justification for the positive aesthetics position. However, if the natural world seems aesthetically good when perceived in its correct categories, this cannot be simply because they are correct; it must be because of the kind of thing nature is and the kinds of categories that are correct for it. This is because the analogous position in art appreciation does not justify positive aesthetics concerning art. Works of art do not necessarily seem aesthetically good when perceived in their correct categories. Yet the only differences between the two views lie in the kind of thing nature is as opposed to art and the kinds of categories that are correct for it as opposed to those for art. Thus, if we are to find a justification for positive aesthetics, these differences must provide the key.

To pursue this line of inquiry it is useful to see more clearly why works of art are not necessarily aesthetically good, even when perceived in their correct categories. Categories of art are established in light of certain facts about works of art and their origins, such as their times and places of creation, their artists' intentions, and their societies' traditions. The determinations of the correct categories for particular works are also a function of such facts. Moreover, the determinations of the aesthetic qualities that particular works or kinds of works have, and thus whether they are aesthetically good or bad, are in part a function of what categories are correct for them. For example, since it is a post-impressionist painting, *The Starry Night* is aesthetically better than it would be were it an expressionist work. Thus, in the case of art, determinations of categories and of their correctness are in general prior to and independent of considerations of aesthetic goodness.[65] This helps to explain why works of art are not necessarily aesthetically good and why there is no positive aesthetics position concerning art. Categories of art are established in such a way that even in their correct categories particular works may be aesthetically bad. However, what if categories of art were established in a different way? In another connection, Walton remarks:

> Take any work of art we can agree is of fourth- or fifth- or tenth-rate quality. It is quite possible that if this work were perceived in some far-fetched set of categories that someone might dream up, it would appear to be first-rate, a masterpiece. Finding such ad hoc categories obviously would require talent and ingenuity on the order of that necessary to produce a masterpiece in the first place. But we can sketch how one might begin searching for them...If the mediocre work suffers from some disturbingly prominent feature that distracts from whatever merits the work has, this feature might be toned down by choosing categories...[such that]...When the work is perceived in the new way the offending feature may be no more distracting than the flatness of a painting is to us...If it needs ingenuity we might devise a set of rules in terms of which the work finds itself in a dilemma and then ingeniously escapes from it, and build these rules into a set of categories.[66]

This passage describes how we might invent categories that would make works of art *seem* aesthetically good. All that is required is that we take the works as given, and create the categories in virtue of them, and with an eye to making them appear aesthetically good.

Walton points out that this procedure would not make a work aesthetically good:

> Surely, however, if there are categories...which would transform a mediocre work into a masterpiece, it does not follow that the work really is a hitherto unrecognized masterpiece. The fact that when perceived in such categories it would appear exciting, ingenious, and so forth, rather than grating, cliché-ridden, pedestrian, does not make it so.[67]

Inventing categories that would seemingly transform mediocre works into masterpieces would not make them masterpieces, because in our world no matter how creatively such categories were produced, they would not be the correct categories for the works in question. However, we can imagine a world different from ours. Imagine one in which "works of art" are not created at all, but rather discovered; and in which "artists" do not have to use their talents and ingenuity to create works of art but rather use them to create categories in which the discovered works appear to be masterpieces. Imagine further that the criterion for categories being correct is that they make the works appear to be masterpieces. In such a world determinations of categories and of their correctness would be dependent upon considerations of aesthetic goodness; and (insofar as the artists accomplish their job) all works of art would in fact be masterpieces. Or to put it another way, all works would be essentially aesthetically good and appropriately appreciated as such. Our imagined world would have positive aesthetics concerning art.

In light of the above it is possible to see the beginnings of a justification for positive aesthetics concerning nature. The idea is that natural objects and landscapes in our world are analogous to the works of art of our imagined world, and scientists in our world are analogous to the artists of our imagined world. That the former is the case is clear enough. Unlike works of art, natural objects and landscapes are not created or produced by humans, but rather "discovered" by them. Only once they are discovered can description, categorization, and theorizing proceed. Thus, natural objects and landscapes are in a sense given, and then the categories of nature are created in virtue of them.

The fact that the natural world is discovered suggests one way in which our scientists are like the artists of our imagined world. Both begin with given objects and use talent and ingenuity to create categories for them. However, the artists of our imagined world determine correctness of categories by the criterion of aesthetic goodness: the correct categories are those that make the given objects appear to be masterpieces. It is not plausible to say that our scientists do simply that, but perhaps they do something quite similar. Aesthetic goodness appears to play some criterialogical role in the scientific enterprise. For example, it is claimed that aesthetic goodness is one of the criteria by which science adjudicates between conflicting descriptions, categorizations, and theories. However, even if this is the case, aesthetic goodness is not the only criterion by which scientists determine correctness of descriptions, categorizations, and theories. Rather the relationship between correctness in science and aesthetic goodness must be both more complex and more contingent.

Perhaps this relationship is somewhat like the following: a more correct categorization in science is one that over time makes the natural world seem more intelligible, more comprehensible to those whose science it is. Our science appeals to certain kinds of qualities to accomplish this. These qualities are ones such as order, regularity, harmony, balance, tension, resolution, and so forth. If our science did not discover, uncover, and/or create such qualities in the natural world and explain the world in terms of them, it would not accomplish its task of making it seem more intelligible to us; rather, it would leave the world incomprehensible, as any of the various world views that we regard as superstition seem to us to leave it. Moreover, these qualities that make the world seem comprehensible to us are also those that we find aesthetically good. Thus, when we experience them in the natural world or experience the natural world in terms of them, we find it aesthetically good. This is not surprising, for qualities such as order, regularity, harmony, balance, tension, and resolution are the kinds of qualities that we find aesthetically good in art. This, I take it, is the idea suggested by those who claim that science and art have similar roots and/or goals; and by those who claim that science is in part an aesthetic endeavor.[68] Beyond this, why there should be such a connection between aesthetic goodness and scientific correctness is not clear. It may be a function of our biology or of our culture; it may be a result of human evolution or an article of humanistic faith; it may simply be a reflection of *our* superstition. However, here it is only necessary to indicate that what we find aesthetically good plays an important role in what our science takes to be correct.

It is now possible to summarize the suggested justification for the positive aesthetics position. The key to the justification lies in the kind of thing nature is as opposed to art and the kinds of categories that are correct for it as opposed to those for art. Art is created, while nature is discovered. The determinations of categories of art and of their correctness are in general prior to and independent of aesthetic considerations, while the determinations of categories of nature and of their correctness are in an important sense dependent upon aesthetic considerations. These two differences are closely related. Since nature is discovered, rather than created, in science, unlike in art, creativity plays its major role in the determinations of categories and of their correctness; and considerations of aesthetic goodness come into play at this creative level. Thus, our science creates categories of nature in part in light of aesthetic goodness and in so doing makes the natural world appear aesthetically good to us. Moreover, the categories created in this way are the correct categories – those that involve appropriate aesthetic appreciation and reveal the aesthetic qualities and value of the objects of that appreciation. Thus, these categories not only make the natural world appear aesthetically good, but in virtue of being correct determine that it is aesthetically good. Or to put the point in a simpler way: the aesthetic situation concerning virgin nature in our world is essentially analogous to that concerning art in our previously imagined world. Our natural objects and landscapes, like its works of art, are discovered and categories are created for them. Our scientists, like its artists, create these categories in virtue of these given objects and, over the long run, with an eye toward aesthetic goodness. And in each world these categories are the correct categories. Thus, our natural objects and landscapes, like its works of art, are essentially aesthetically good. The result is that we have positive aesthetics concerning nature as it would have positive aesthetics concerning its art.

If this line of thought yields a justification for the positive aesthetics position, it does so with an important qualification. The justification is not of the same kind as those rejected earlier. For example, the divine justification, if adequate, would support the claim not simply that virgin nature is aesthetically good, but also that it always has been and always will be – whether anyone ever notices or not. The justification developed here, however, regards the aesthetic appreciation of nature as significantly informed by science and positive aesthetics as intimately related to the development of science. Consequently, although aesthetic appreciation of nature is perhaps informed by whatever world view is available, it seems that, outside the temporal and spatial boundaries of the scientific world view, it is not informed by science. Thus, positive aesthetics may not be a justifiable position outside these boundaries. This justification is, as it were, within the scope of the scientific world view. The nature and extent of positive aesthetics that is justified, therefore, seemingly depends upon interpretations of science. In light of various views about science, positive aesthetics might be, for example, absolute, culturally relative, or paradigm relative.[69]

To some this potential limitation of the positive aesthetic position may seem unfortunate, but it nonetheless adds plausibility to the justification itself. It helps to explain the close correlation between the development of natural science and the development of the aesthetic appreciation of nature noted in the two previous sections of this chapter. Moreover, if we construe the development of science as the attempt to make the natural world seem more and more comprehensible to us, by continual self-revision, whenever it appears less then fully comprehensible, then we also have an explanation of the growth of positive aesthetics. If comprehensibility is in part a function of qualities such as order, regularity, harmony, and balance, which we find aesthetically good, then the development of science and its continual self-revision constitutes a movement toward the aesthetically good. Or perhaps less misleadingly, it constitutes a movement that puts the natural world in an increasingly favorable aesthetic light. Such a movement seems evident in general and also at more specific levels.The positive aesthetic appreciation of previously abhorred landscapes, such as mountains and jungles, seems to have followed developments in geology and geography.[70] Likewise, the positive aesthetic appreciation of previously abhorred life forms, such as insects and reptiles, seems to have followed developments in biology.[71] In retrospect, many of the advances in natural science can be viewed as heralding a corresponding advance in positive aesthetics. This in itself gives support to the justification suggested in this section.

In a similar way, if this justification is correct, another connection noted in preceding sections falls into place – the connection between the development of positive aesthetics as a position concerning the natural world in general and the birth and growth of ecology. This is understandable in light of the fact that ecology not only is in certain respects all encompassing, but also puts considerable emphasis on qualities such as unity, harmony, and balance – ones we find particularly aesthetically good. I indicated that the positive aesthetics position is most evident in the writings of contemporary individuals especially concerned with ecology and ecological issues. Perhaps this position is not simply to be justified as a scientific aesthetic but is indeed what one individual terms the "ecological esthetic."[72] The position has seemingly come into its own with the development of ecology and seemingly continues to grow in light of it. Some of the individuals noted earlier suggest this. For example, Rolston writes:

> We do not live in Eden, yet the trend is there, as ecological advance increasingly finds in the natural given stability, beauty, and integrity, and we are henceforth as willing to open our concepts to reformation by the world as to prejudge the natural order.[73]

Perhaps someday, at least concerning the natural world, we may all agree with Constable, saying "I never saw an ugly thing in my life."[74]

Notes

1 For an enumeration of uses of "nature" in the appeal to nature for norms of art, especially with reference to the eighteenth century, see Arthur Lovejoy, " 'Nature' as Aesthetic Norm," *Modern Language Notes*, 1927, vol. 42, pp. 444–50. For a brief discussion of the eighteenth-century point of view, see Harold Osborne, "The Use of Nature in Art," *British Journal of Aesthetics*, 1962, vol. 2, pp. 318–27.

2 The comment is quoted and dated in Andrew Forge, "Art/Nature," *Philosophy and the Arts: Royal Institute of Philosophy Lectures*, London, Macmillan, 1973, vol. 6, pp. 231. Forge remarks: "The position that he was affirming was to be an ingredient in almost all the most vital painting of the next sixty years."

3 John Ruskin, *The Elements of Drawing* [1857], New York, Dover, 1971, pp. 128–9. The passage continues:

> The rock on which the effect of a mountain scene especially depends is always precisely that which the roadmaker blasts or the landlord quarries; and the spot of green which Nature left with a special purpose by her dark forest sides, and finished with her most delicate grasses, is always that which the farmer ploughs or builds upon. But the clouds...cannot be quarried nor built over, and they are always therefore gloriously arranged;...they all move and burn together in a marvellous harmony; not a cloud of them is out of its appointed place, or fails of its part.

4 The two characterizations are from respectively Lewis Mumford, *The Brown Decades: A Study of the Arts in America, 1865–1895*, New York, Harcourt, Brace and Co., 1931, p. 78 and Stewart L. Udall, *The Quiet Crisis*, New York, Holt, Rinehart and Winston, 1963, p. 82. Both are quoted by David Lowenthal in his introduction to George Perkins Marsh, *Man and Nature* [1864], Cambridge, Harvard University Press, 1965, pp. ix and xxii.

5 Ibid., pp. 8–9.

6 William Morris, *Art and the Beauty of the Earth: A Lecture Delivered at Burslem Town Hall on October 13, 1881*, London, Longmans and Company, 1898, p. 24. For Morris's thoughts on landscape reform, see William Morris, *The Beauty of Life: An Address Delivered at the Town Hall, Birmingham, in 1880*, London, Brethan Press, 1974, especially pp. 10–18.

7 John Muir, "The Wild Parks and Forest Reservations of the West," in *Our National Parks*, Boston, Houghton Mifflin, 1916, pp. 6–7.

8 David Lowenthal, "The American Scene," *Geographical Review*, 1968, vol. 58, p. 81.

9 Leonard A. Fels, "Aesthetic Decision-Making and Human Ecology," in *Proceedings of the VIIth International Congress of Aesthetics*, Bucharest, Editura Academiei Republic Socialist Romania, 1977, p. 369.

10 Lilly-Marlene Russow, "Why Do Species Matter?," *Environmental Ethics*, 1981, vol. 3. p. 109.

11 Kenneth H. Simonsen, "The Value of Wildness," *Environmental Ethics*, 1981, vol. 3, p. 263.

12 Holmes Rolston, III, "Can and Ought We to Follow Nature?," *Environmental Ethics*, 1979, vol. 1, pp. 23–4. Later in the article Rolston qualifies this claim somewhat: "My concept of the good is not coextensive with the natural, but it does greatly overlap it; and I find my estimate steadily enlarging that overlap" (p. 28). See also Holmes Rolston, III, "Is There an Ecological Ethic?," *Ethics*, 1975, vol. 85, pp. 102–3.

13 Joseph W. Meeker, *The Comedy of Survival: Studies in Literary Ecology*, New York, Charles Scribner's Sons, 1974, p. 129. Meeker, following Konrad Lorenz, holds that in both animals and man "wild" characteristics are beautiful and "virtually all characters we perceive as specifically ugly are genuine domestication effects" (see pp. 121–4). The source is Konrad Lorenz, *Studies in Animal and Human Behavior*, London, Methuen, 1971. See also Yi-Fu Tuan, "Visual Blight: Exercises in Interpretation," *Visual Blight in America*, Commission on College Geography Resource Paper no. 23, Washington, D.C., Association of American Geographers, 1973, p. 27. Tuan attributes the root idea to Susanne Langer, *Feeling and Form*, New York, Charles Scribner's Sons, 1953.

14 Meeker, op. cit., p. 136.

15 Robert Elliot, "Faking Nature," *Inquiry*, 1982, vol. 25, p. 84. Elliot does not take himself to be discussing aesthetic appreciation and evaluation in this article, but rather what he terms "environmental evaluation," which he holds is differentiated from

aesthetic evaluation by "the judgmental element" in the latter (see p. 90). As I indicate in the next section, I believe that the argument for making this differentiation is not adequate and that he is mistaken in thinking that the relevant judgments and evaluations are not genuinely aesthetic. Consequently I consider his view a moderate version of positive aesthetics.

16 Ibid., pp. 86–7.

17 Ibid., p. 87.

18 Elliot's argument is similar to that of Nelson Goodman in Chapter 3 of *Languages of Art*, Indianapolis, Bobbs-Merrill, 1968.

19 Aarne Kinnunen, "Luonnonestetiikka," in Aarne Kinnunen and Yrjo Sepanmaa (eds) *Ymparistoestetiikka*, Helsinki, Gaudeamus, 1981, p. 49. The quote is translated from the original Finnish by Anja Sahuri. Kinnunen's idea of positive aesthetics is also discussed in Yrjo Sepanmaa, "Ymparistoestetiikka," in Yrjo Varpio (ed.) *Taiteentutkimaksen Perusteet*, Helsinki, Werner Soderstrom Osakeyhtio, 1982, pp. 33–46. I am grateful to Yrjo Sepanmaa for his contribution to my thinking about the topic of this chapter. Many of the relevant ideas and sources came to my attention during a series of fruitful discussions with him throughout 1982. I also thank J. Baird Callicott and Mark Sagoff for helpful comments.

20 Elliot, op. cit., p. 90. All the short quotes from Elliot in this section are also from p. 90.

21 Osborne, op. cit., p. 325. See also Don Mannison, "A Prolegomenon To a Human Chauvinistic Aesthetic," in D.S. Mannison, M.A. McRobbie, R. Routley (eds) *Environmental Philosophy*, Canberra, Australia National University, 1980, pp. 212–16.

22 This view is endorsed by theories of aesthetic appreciation as diverse as those held by, for example, George Dickie and Jerome Stolnitz. Stolnitz says that "the aesthetic attitude can be adopted toward 'any object of awareness whatsoever.'" See Jerome Stolnitz, *Aesthetics and the Philosophy of Art Criticism*, New York, Houghton Mifflin, 1960, pp. 40–2. Dickie holds that limiting an aesthetic theory in virtue of the "appreciatability" of objects "is vacuous, since it is unlikely that any object would lack some quality which is appreciatable." See George Dickie, "A Response to Cohen: The Actuality of Art," in George Dickie and R. J. Sclafani (eds) *Aesthetics: A Critical Anthology*, New York, St. Martin's, 1977, p. 200.

23 Paul Ziff, "Anything Viewed," in Esa Saarinen, Risto Hilpinen, Ilkka Niiniluoto, and Merrill Provence Hintikka (eds) *Essays in Honour of Jaakko Hintikka*, Dordrecht, Reidel, 1979, p. 285.

24 Neil Evernden, "The Ambiguous Landscape," *Geographical Review*, 1981, vol. 71, p. 155.

25 Ziff, op. cit., p. 293.

26 Elliot's claim is actually that the "apparently integral part of aesthetic evaluation depends on viewing the aesthetic object as an intentional object, as an artifact, as something that is shaped by the purposes and the design of its author," not that it depends on the object having these features. This further weakens the argument. Given this, all that is required for environmental evaluations to involve the "apparently integral part of aesthetic evaluation" and thus be aesthetic is for the appreciator to either mistakenly or intentionally view the natural world as having these features. Although it is certainly possible for an appreciator to view nature in this way, I do not develop this weakness, for I agree with the spirit of many of Elliot's remarks that suggest that this would be an incorrect, inappropriate, or at least misleading way to view any natural environment. I discuss some of the problematic aspects of viewing the natural world in such ways in "Appreciation and the Natural Environment," *Journal of Aesthetics and Art Criticism*, 1979, vol. 37, pp. 267–75 (reproduced in this volume, Chapter 4).

27 Ziff, op. cit., pp. 286–7.

28 The most important sources are Edmund Burke, *A Philosophical Enquiry into the Origin of our Ideas of the Sublime and Beautiful* [1757], J. T. Boulton (ed.), London, Routledge and Kegan Paul, 1958, and Immanuel Kant, *Observations on the Feeling of the Beautiful and Sublime* [1763], trans. J. T. Goldthwait, Berkeley, California, 1965 and *Critique of Judgment* [1790], trans. J. H. Bernard, New York, Hafner, 1968.
29 Simonsen, op. cit., p. 261.
30 Ibid., p. 263. All the short quotes from Simonsen in the remainder of this section are also from p. 263.
31 Nelson Potter, "Aesthetic Value in Nature and in the Arts," in Hugh Curtler (ed.) *What is Art?*, New York, Haven Publications, 1983, pp. 142–3.
32 Simonsen, op. cit., p. 263.
33 Potter, op. cit., pp. 142–3.
34 Ibid., p. 143.
35 Ibid., p. 143.
36 For example, John Hick contends that so far as Judaic-Christian theism is concerned certain solutions to the problem of evil "have to be ruled out." Within this context, he holds, solutions which fall within my first category are "impossible" and those which fall within my second category "have abandoned the basic premise of Hebrew-Christian Monotheism." His entire discussion of traditional and contemporary attempts to solve the problem concerns theodicy. See John Hick, *Philosophy of Religion*, Englewood Cliffs, Prentice-Hall, 1963, pp. 44–7.
37 A classic statement of this view is Lynn White, Jr., "The Historical Roots of Our Ecologic Crisis," *Science*, 1967, vol. 155, pp. 1203–7.
38 Ronald Rees, "The Taste for Mountain Scenery," *History Today*, 1975, vol. 25, p. 307.
39 Ibid., p. 308; quoted from Thomas Burnet, *Sacred Theory of the Earth* [1684].
40 Victor Romanenko, "The Beauty of Nature," trans. Bernard Isaacs, in S. Mozhnyagun (ed.) *Problems of Modern Aesthetics*, Moscow, Progress Publishers, 1969, p. 143.
41 Ibid., p. 139.
42 Marjorie Hope Nicolson, *Mountain Gloom and Mountain Glory*, Ithaca, Cornell University Press, 1959. It is generally accepted that art, especially landscape painting, has played an important role in the development of nature appreciation. See, for example, Alfred Biese, *The Development of the Feeling for Nature in the Middle Ages and Modern Times*, New York, Burt Franklin, 1905, E. W. Manwaring, *Italian Landscape in XVIII Century England*, New York, Oxford University Press, 1925, or Christopher Hussey, *The Picturesque*, London, G. P. Putnam's, 1927. To understand the complexity of the subtle interplay between art, science, and nature appreciation, in particular in relation to North American environmental attitudes, see Eugene Hargrove, "The Historical Foundations of American Environmental Attitudes," *Environmental Ethics*, 1979, vol. 1, pp. 209–40.
43 Romanenko, op. cit., p. 143.
44 Ibid., p. 141.
45 John Constable, from his last public lecture, delivered at the Royal Institution of Great Britain in 1836. Quoted in Ronald Rees, "John Constable and the Art of Geography," *Geographical Review*, 1976, vol. 66, p. 59.
46 Ibid., pp. 59–61; quoted from John Constable, in C. R. Leslie (ed.) *Memoirs of John Constable*, London, Phaidon, 1951, p. 272. On Constable and science, see also E. H. Gombrich, *Art and Illusion*, Princeton, Princeton University Press, 1961, especially part I.
47 Rees, "John Constable and the Art of Geography," op. cit., p. 59. On the role of science in Ruskin's thought, see Denis E. Cosgrove, "John Ruskin and the Geographical Imagination," *Geographical Review*, 1979, vol. 69, pp. 43–62.

48 With the exception of the second and the sixth, these are respectively, the titles of the following articles: Fels, op. cit., Rolston, "Can and Ought We to Follow Nature?," op. cit., Rolston, "Is there an Ecological Ethics?," op. cit., and Simonsen, op. cit. The second and sixth are chapter and section titles from Meeker, op. cit.
49 Kinnunen, op. cit., pp. 47–8.
50 Elliot, op. cit., p. 91.
51 Rolston, "Can and Ought We to Follow Nature?," op. cit., p. 23.
52 Rolston, "Is There an Ecological Ethic?," op. cit., p. 107. Rolston is quoting the last three words from Aldo Leopold, "The Land Ethic," *A Sand County Almanac*, New York, Oxford University Press, 1949, p. 223. In these articles Rolston is more concerned with moral than with aesthetic value.
53 Meeker, op. cit., pp. 124–5.
54 Biese, op. cit., p. 357.
55 Val Routley, "Critical Notice of John Passmore, *Man's Responsibility for Nature*," *Australasian Journal of Philosophy*, 1975, vol. 53, p. 183.
56 Elliot, op. cit., p. 91.
57 Rolston, "Is there an Ecological Ethic?," op. cit., pp. 100–1.
58 Routley, op. cit., p. 183.
59 Kendall Walton, "Categories of Art," *Philosophical Review*, 1970, vol. 79, p. 364.
60 Paul Ziff, "Reasons in Art Criticism," in W. E. Kennick (ed.) *Art and Philosophy: Readings in Aesthetics*, New York, St. Martin's Press, 1964, p. 620.
61 Walton, op. cit., p. 338–9.
62 Ziff, "Anything Viewed," op. cit., p. 291.
63 Ronald Hepburn, "Aesthetic Appreciation of Nature," in Harold Osborne (ed.) *Aesthetics in the Modern World*, London, Thames and Hudson, 1968, p. 55.
64 See my "Appreciation and the Natural Environment," op. cit., (reproduced in this volume, Chapter 4) and especially "Nature, Aesthetic Judgment, and Objectivity," *Journal of Aesthetics and Art Criticism*, 1981, vol. 40, pp. 15–27 (reproduced in this volume, Chapter 5). In the latter chapter, as here, I utilize some of the terminology introduced by Walton in "Categories of Art," op. cit. In that article Walton does not apply his category approach to the aesthetic appreciation of nature and in fact expresses doubts about the possibility of doing so. However, in subsequent conversations, he has agreed that it is plausible to understand the aesthetic appreciation of nature in terms of natural categories, as I argue in "Nature, Aesthetic Judgment, and Objectivity."
65 In Walton's account the two are not necessarily independent, for he holds that the fact, if it is one, that a work is more interesting or pleasing aesthetically when perceived in a given category than it is when perceived in alternative ways is a circumstance that counts "toward its being correct to perceive a work" in that category (op. cit., p. 357). I think this is mistaken; see "Nature, Aesthetic Judgment, and Objectivity," op. cit. (reproduced in this volume, Chapter 5).
66 Walton, op. cit., pp. 359–60.
67 Ibid., p. 360.
68 These and related claims are illustrated and discussed in Judith Wechsler (ed.) *Aesthetics in Science*, Cambridge, Massachusetts, The MIT Press, 1978. The theme of the collection, ibid., pp. 6, 1, is:

> The search in science for models that illuminate nature seems to parallel certain crucial processes in art…When scientists…reflect on their work, the development of concepts, and the theories that expound them, it is evident that intuition and aesthetics guide their sense of "this is how it has to be," their sense of rightness.

69 Another possible qualification is that the justification offered here supports positive aesthetic appreciation of kinds of natural things but not of natural particulars. I doubt this because, given the role of aesthetic goodness in scientific description, categorization, and theorizing, I suspect that scientific knowledge as a whole is aesthetically imbued such that our appreciation of particulars is as enhanced as is that of kinds.

70 Positive aesthetic appreciation of unfamiliar landscapes such as volcanic mountains and tropical jungles was probably greatly enhanced by the scientific discoveries and writings of Alexander von Humboldt. It is noteworthy that his *Ansichter der Natur* [1808] went through three editions and was translated into nearly every European language by the time of his death in 1859. For a brief discussion of Humboldt's influence on the aesthetics of nature, see Edmunds Bunkse, "Humboldt and an Aesthetic Tradition in Geography," *Geographical Review*, 1981, vol. 71, pp. 127–46.

71 Concerning positive aesthetic appreciation of life, Darwin is particularly important. It is sometimes contended that Darwin's work had an adverse effect on the appreciation of nature. For example, Rolston notes: "After Darwin (through misunderstanding him, perhaps), the world of design collapsed, and nature, for all its law, seemed random, accidental, chaotic, blind, crude." "Is There an Ecological Ethic?," op. cit., p. 107). I agree that this may involve misunderstanding and think there is insight in Romanenko's remark that "everything, which before the days of Darwin had borne the stamp of 'divine origin', the beauty of nature included, was passed down to the earth from heaven" (op. cit., p. 141). For example, prior to Darwin's demonstration of the transmutability of species these important categories of nature could be viewed as God-given. Darwin's work shows how they are the result of the human endeavors of scientific discovery, generalization, and decision. In addition, the expurgation of divine design effected by Darwin's work undercuts the grounds for certain kinds of negative aesthetic judgments about certain aspects of nature. Darwin's view that evolution has no overall direction and that all life forms must be accounted for in the same natural terms, gives no basis for viewing some as aesthetically inferior to others. As Stephen Jay Gould remarks, in light of the view that evolution "does not lead inevitably to higher things…The 'degeneracy' of a parasite is as perfect as the gait of a gazelle." See Stephen Jay Gould, *Ever Since Darwin*, New York, W. W. Norton, 1977, p. 13.

72 Meeker, op. cit., p. 119.

73 Rolston, "Is there an Ecological Ethic?," op. cit., p. 108.

74 It may be thought that a conclusive objection to this justification is simply that it tends to justify positive aesthetics, a position that is somewhat implausible. In short, the objection is that since there is much in the natural world that we do not find aesthetically good, any justification of the position must be incorrect. I agree that there is much in the natural world that appears to many of us not to be aesthetically good. However, this fact itself does not constitute a conclusive objection, for the justification provides the means of showing how the fact is consistent with the positive aesthetics position. First, as suggested by Rolston's remark that "we do not live in Eden, yet the trend is there," it is understandable if at the present we do not find all the natural world essentially beautiful. If our positive aesthetic appreciation of nature follows and is dependent upon the development of science, then it is to be expected that at this point in time there is much in the natural world that we do not yet find aesthetically good. Moreover, this is especially to be expected if the most relevant "science" is ecology, for it is not only a comparatively recent development but has yet to achieve the status of a mature science. For useful discussion, see Robert T. McIntosh, "The Background and Some Current Problems of Theoretical Ecology," in Esa Saarinen (ed.) *Conceptual Issues in Ecology*, Dordrecht, Reidel, 1982, pp. 1–61. Second, although the correct categories for the aesthetic appreciation of the natural world are natural categories, there are other categories in terms of which we do, as a

matter of fact, occasionally perceive nature. When nature is perceived in such categories, there is, according to this justification, no reason why positive aesthetic appreciation should result. Indeed, when so perceived, much of the natural world may appear aesthetically second-rate – bland, dull, insipid, incoherent, chaotic, and the like. Whether or not certain of these other categories should also be accepted as correct categories for the natural world is another issue. Although this cannot be pursued here, I, as suggested previously, am inclined to think that they should not be so accepted – at least not for virgin nature (and not within the scientific world view).

7

APPRECIATING ART AND APPRECIATING NATURE

The concept of appreciation

The concept of appreciation is common to both art appreciation and nature appreciation. However, it is usually not examined in the relevant theoretical work. Writings on appreciating art by art critics and art historians seldom touch on it. Nature literature may exemplify it but typically does not discuss it. Investigations of aesthetic appreciation by aestheticians dwell on the nature of the aesthetic and have little to say about appreciation. That the concept is not discussed is a pity, for it is central both to philosophical aesthetics and to our day-to-day dealings with such matters. Not only are the notions of art appreciation and nature appreciation in common usage, but we move with ease from the appreciation of landscapes to that of landscape paintings, from appreciating Van Gogh's *The Starry Night* (1889) to appreciating the starry heavens above. Yet the nature of appreciation is far from clear and what is involved in each of these two central cases – appreciating art and appreciating nature – remains obscure.

Thus, some clarification of appreciation is useful. To achieve it, since the topic of this chapter is the appreciation of both art and nature, it is appropriate to consider a philosophical tradition that in its infancy thought nature at least as significant as art as an object of aesthetic appreciation. This tradition, which was introduced and briefly considered in Chapters 1 and 2, ties appreciation to notions such as disinterestedness. Although this position like most others in philosophical aesthetics is more concerned with the aesthetic than with appreciation, it yet provides insight into the nature of appreciation. Equally important, it points the way to a flaw in much philosophical thinking about appreciation – a flaw that apparently stems from its myopic focus on the aesthetic. Both the insight and the flaw can be clarified, adequately for present purposes, by considering the recent version of the position presented by Jerome Stolnitz and the attack on it by George Dickie.[1] Stolnitz's elaboration of "the aesthetic attitude" illustrates the insight; Dickie's elaboration of "the myth of the aesthetic attitude" brings out the flaw.

Stolnitz's version of the disinterestedness position focuses on a special attitude, the aesthetic attitude, defined as "disinterested and sympathetic attention to and contemplation of any object of awareness whatever, for its own sake alone."[2] True to the tradition, Stolnitz does not directly discuss aesthetic *appreciation*. However, it is to be treated, it may be assumed, as related notions such as "aesthetic experience" which are "defined by reference to" the aesthetic attitude.[3] Thus, aesthetic appreciation would be defined as the total appreciation engaged in while this attitude is being taken. Consequently, insights into the nature of appreciation are provided in Stolnitz's remarks about taking the aesthetic attitude.

The most fundamental of these insights about aesthetic appreciation concerns what may be termed its scope. Although at points in its history the disinterestedness tradition apparently judged certain kinds of objects to be closed to disinterested attention and therefore essentially nonaesthetic, the tradition by and large emphasizes a broad scope for appreciation. This is cited by Stolnitz as one of two major reasons for preferring this line of approach to others. In Stolnitz's version of the position the scope is in fact "limitless" in that the "aesthetic attitude can be adopted toward 'any object of awareness whatever.' "[4] The recognition of such a scope for aesthetic appreciation is particularly important for understanding nature appreciation, for nature, noted for diversity, comes in all shapes and sizes and all types and kinds, many seemingly not tailor-made for appreciation as are paradigmatic works of art.

A second significant dimension of appreciation can be clarified by noting how it differs from some other notions utilized in the tradition and in fact contained in Stolnitz's definition – notions such as contemplation and awareness. Here Stolnitz's appeal to the concept of an attitude is helpful, although often it is simply misleading. As Stolnitz emphasizes, attitudes are directive, they organize, orientate, and guide. This directive nature "prepares us to *respond*" and the responsiveness of appreciation separates it from passive states such as contemplation or awareness.[5] Indeed, Stolnitz thinks it "safe" to use the word "contemplation" only after stressing that taking the aesthetic attitude is an "alert and vigorous" business in which we focus "discriminating attention" upon the object, " 'key up' our capacities of imagination and emotion to respond to it," and engage in a range of emotional, cognitive, and physical "activity."[6] In this way appreciation is severed from what Stolnitz describes as the "blank, cow-like stare" often associated with contemplation, with disinterestedness, and with the aesthetic itself.[7] Rather it is aligned with the slogan Stolnitz adopts from a psychologist: "Appreciation...is awareness, alertness, animation."[8]

The active nature of appreciation helps to illuminate the flaw that mars the tradition's treatment of it, a flaw central to Dickie's attack. The problem develops because, unlike either a blank, cow-like stare that needs no guidance or serene contemplation that needs only a little, active, responsive appreciation requires considerable guidance. Stolnitz relies on attitudes here, which since they organize, orientate, and guide, provide by means of the aesthetic attitude, a *general aesthetic criterion* for the guidance of appreciation. Dickie attacks this as the first way the disinterestedness tradition "misleads aesthetic theory": the way it sets "the limits of aesthetic relevance."[9] This may be called the question of aesthetic relevance, the question of how to guide appreciation or, more precisely, how to determine what is relevant to the appropriate appreciation of particular objects. It is not only the issue upon which the disinterestedness treatment of appreciation stumbles, it is a key issue in any such treatment.

The tradition's problems with aesthetic relevance stem from a tension among its essential elements. The tension is present throughout the tradition but is especially evident in Stolnitz's definition which requires attention to be both disinterested and sympathetic. Disinterestedness pulls toward the general criterion of aesthetic relevance, sympathy in the other direction. The tension is illuminated by the responsive nature of appreciation. By contrast, the blank cow-like stare can be both disinterested and sympathetic. What does it matter? It does not respond. But since appreciation is responsive, to sympathetically respond it must, as Stolnitz says, "accept the object 'on its own terms,' " "follow the lead of the object and respond in concert with it" – only in this way can we "relish its individual quality."[10] On the other hand, disinterestedness requires an experience that "at its best, seems to isolate both us and the object from the flow of experience," one in which the object "is divorced from its interrelations with other things."[11] The question is what "other things" then – in particular, what, as Stolnitz puts it, "thoughts or images or bits of knowledge which are not present within the object itself" – are relevant to its appreciation?[12] The tradition's answer – the general criterion of aesthetic relevance – is that it depends on whether any such thought, image, or bit of knowledge is "aesthetic," which in turn "depends on whether it is compatible with the attitude of 'disinterested attention.' "[13] The upshot is that much that might enhance the appreciation of an object – help us to "relish its individual quality" – is condemned as nonaesthetic and therefore as irrelevant to aesthetic appreciation.

Thus, the elaborations of disinterestedness and sympathy place these two concepts at odds. To make the conflict perfectly clear only requires the right kinds of objects. In responsive appreciation we must "follow the lead of the object" yet isolate and divorce it and ourselves from "its interrelations with other things," unless such things are compatible with "disinterested attention." But what if an object does not lead in that direction, what if it resists being isolated and divorced from its interrelations? Indeed, although some works of art may be "aesthetic" in the sense that they readily yield to being – in fact are explicitly created to be – so isolated and divorced, many other works and most nonart, and nature in particular, are precisely not "aesthetic" in this limiting sense. Thus, to respond to such objects as if they were is not to follow their lead, not to be sympathetic to them, not to appreciate them. Does this mean that these "nonaesthetic" objects that cling to their interrelations with other things are therefore closed to "disinterested attention"? This cannot be, for, as noted, at least at this late date in the tradition, the aesthetic attitude is limitless, it can be taken toward "any object of awareness whatever." It becomes clear that something must go.

As is typical in such cases, however, the reaction is an overreaction. It is clear that something must go and what has gone is most of the disinterestedness tradition along with its insight concerning appreciation. The isolation of the appreciator and the object and the divorcing of the latter from its interrelations, all seemingly required by disinterestedness, are taken, with some justification, to after all reduce appreciation to the notorious cow-like stare. With this reductio in mind, critics lose sight of the rich, expansive, and responsive notion of appreciation contained in the tradition. Indeed, some reactions are more extreme. Dickie, for example, overreacting to his own critique, seemingly rejects the very concept of the aesthetic, stating that "there is no reason to think that there is a special kind of aesthetic consciousness, attention, or perception." He adds: "Similarly, I do not think there is any reason to think that there is a special kind of aesthetic appreciation."[14] This may be to throw out the baby with the bath water, but it has a point: the problem with the disinterestedness tradition, as with philosophical aesthetics in general, is that too much attention is paid to the concept of the aesthetic and too little to that of appreciation. The attempt to accommodate the former warps disinterestedness and therefore appreciation itself into a restrictive and isolating state, caricaturizable as the cow-like stare. A shift of emphasis to the latter yields a different picture.

Interesting enough, such a shift of emphasis, together with its happy consequences, is evident within the disinterestedness tradition itself. For example, once Stolnitz moves from philosophical analysis of the aesthetic attitude to consideration of art appreciation, the story changes dramatically. Instead of the strict application of the general criterion of aesthetic relevance, we find that concerning, for instance, the issue of relevant knowledge: "We need not, however, condemn all 'knowledge about' as aesthetically irrelevant… 'Knowledge about' is relevant under three conditions: when it does not weaken or destroy aesthetic attention to the object, when it pertains to the meaning and expressiveness of the object, and when it enhances the quality and significance of one's immediate aesthetic response to the object."[15] Note that only the first condition accommodates the aesthetic; the latter two aim at enhancing the appreciation of the object. With the emphasis thus shifted to appreciation, sympathy outweighs disinterestedness, and we truly "follow the lead of the object." Moreover, it is now not the blind leading the blind, for the aestheticizing cow-like stare gives way to appreciation not only responsive to the object but informed by knowledge about it.

The shift of emphasis to appreciation is followed up by Paul Ziff. Ziff's treatment is informative in that it retains the insights of the disinterestedness position without embracing its flaws.[16] As noted in Chapters 4 and 6, the essence of his account is the notion of an "act of aspection," the way of attending to an object that in part constitutes its appropriate appreciation.[17] Ziff argues that different acts of aspection are appropriate in the appreciation of, for example, works of art of different kinds, styles, and schools. Thus, knowledge of a work's history and nature dictates the proper acts of aspection: appreciation is a set of activities not only responsive to the object but incorporating knowledge of it as an essential component. Ziff further argues not simply that "anything that can be viewed is a fit object for aesthetic attention," but that "anything viewed makes demands."[18] That objects of appreciation make demands means that following the lead of the object rules out the possibility of anything like a general criterion of aesthetic relevance. Since objects of appreciation obviously differ a great deal, so does what is relevant to and involved in their appreciation: "As the character of the objects attended to vary, the character of the actions, the conditions, and the requisite qualities, skills, and capacities of the person may also have to vary, if attention to the objects is to be aesthetically worthwhile."[19] Thus, the appreciation of each of, to use some of Ziff's examples, a work of art by Leonardo Da Vinci, a *Barry McKenzie* comic strip, an alligator basking in the sun, or a pile of dried dung requires engaging in different acts of aspection, using different capacities and skills, and knowing different things. The general criterion of aesthetic relevance is replaced by object-given indications of appreciative relevance.

Stolnitz's and Ziff's remarks on appreciation demonstrate that although philosophical aesthetics has relatively little to say about the concept, it is yet possible to derive from its investigation of the aesthetic some useful observations. Ironically it is precisely in drawing back from the tradition's obsession with the aesthetic that the concept of appreciation is brought into focus. The obsession with the aesthetic inhibits a proper understanding of appreciation by pulling in the direction of a passive state of limited scope, restricted by a general criterion of aesthetic relevance, and comparable to a blank, cow-like stare. By contrast, the concept of appreciation that can be coaxed from philosophical aesthetics, seemingly almost against its will, reveals appreciation as engaged mental and physical activity applicable to any object whatever, exceedingly responsive to that object, and guided almost exclusively by its nature.

We may conclude this discussion of appreciation by asking, in light of the object-orientated nature of appreciation, what remains of the idea that the tradition attempts to capture with concepts such as disinterestedness? What becomes of the isolating attention intended to divorce both the object and the appreciator from all that is irrelevant to the object's appreciation? In fact there is in the object-orientated notion a significant residue of this idea. It is that to follow the lead of the object and be guided by it is to be "object-ively" guided. This sense of objective is the most basic: it concerns the object and its properties and is opposed to the subjective in the sense of concerning the subject and its properties. Appreciating objectively in this sense is appreciating the object as and for what it is and as and for having the properties it has. It is in opposition to appreciating subjectively in which the subject – the appreciator – and its properties are in some way imposed on the object, or, more generally, something other than the object is imposed on it.

Thus, the insight of the disinterestedness part of the tradition is that, insofar as appreciation is disinterested, it is objective. And when it fails to be, it is in a corresponding sense subjective and involves viewing the object as something it is not or as having properties it does not have. Thus, appreciation is isolating only in the sense that both object and appreciator must be divorced from that which is not true of the object. In short, it is the false that is incompatible with disinterested attention and therefore irrelevant to aesthetic appreciation. The insight contained in the passive, blank, cow-like stare is that it limits itself to the dull, objective truth. Moreover, in that disinterestedness constitutes part of the analysis of the aesthetic, aesthetic appreciation is therefore no more *nor less* than appreciation in this sense objective.[20] To appreciate something aesthetically is to appreciate it as and for, to use the phrase Butler coined in another context, "what it is, and not another thing."[21] No wonder everything is open to aesthetic appreciation.

The notion of object-orientated aesthetic appreciation that thus emerges from philosophical aesthetics is precisely the kind of concept required for a fruitful investigation of appreciating art and appreciating nature. Given the "limitless" scope that it grants appreciation, such a notion is especially useful in understanding the appreciation of diverse kinds of things, such as works of art and natural objects. Moreover, the concept of object-orientated appreciation facilitates the constructive comparison of different kinds of appreciation, of art appreciation and nature appreciation, without the assimilation of one to the other. Without a concept of appreciation that allows, indeed requires, appreciative activity to respond directly to and to vary according to the nature of different kinds of objects, there is a danger of all appreciation being assimilated to one model – typically that of the appreciation of the most conventional kind of art. However, although it must not be taken as the model for all other forms of appreciation, the appreciation of conventional and therefore paradigmatic works of art is yet the proper starting point for any investigation of different kinds of appreciation.

Appreciating art: design appreciation

What is involved in the appreciation of paradigmatic works of art or, more precisely, what is paradigmatic art appreciation? Given the object-orientated nature of appreciation, there must be diverse kinds of art appreciation, each requiring engaging in different physical and mental acts of aspection, using different capacities and skills, and knowing different things. Indeed, they may have little in common, not even the utilization of any one sense modality. Nonetheless, the question is what is central to all of these many diverse kinds of appreciation – what in general is significant in art appreciation? Seemingly the only possible answer is the appreciation of design: paradigmatic art appreciation must be at least appreciation of a thing as something designed and therefore as something that is the creation of a designer. To put it another way, appreciation of art qua art must be appreciation qua creation of an artist.

That art appreciation is artist or designer centered seems an obvious point, but it is not always fully appreciated in philosophical aesthetics. To some extent the prominence of the disinterestedness tradition is responsible. As noted, when developed in certain ways, the notion of disinterestedness requires that the object of appreciation be isolated and divorced from its interrelationships with other things, its appreciation being strictly constrained by the general criterion of aesthetic relevance. The result is a purified aesthetic object, divorced from its own history, even from the fact that it is the product of a designer. Moreover, in recent philosophy of art criticism this perverse offspring of disinterestedness gives comfort to and joins with anti-intentionalism in an attempt to ban almost any knowledge of the artist from the appreciation of his or her creations.[22] Thus, the disinterestedness tradition, with a little assistance, completely obscures the designer-centered nature of paradigmatic art appreciation.

This line of thought, however, stems from the same mishandling of disinterestedness that obscures the object-orientated nature of appreciation. Consider the fact that even if a work is explicitly designed by its creator to be an isolated, pure aesthetic object, narrowly "aesthetic" in the sense required by the general criterion of aesthetic relevance, it is nonetheless still *designed by its creator* to be that way. Thus, appreciation that to any extent follows the lead of the object, although it may appropriately ignore many other facts about such an object, simply cannot ignore the central fact that the object is designed to be the way it is – that its being designed is the essence of its being "what it is and not another thing." Consequently, in that appreciation is objective in the sense noted in the first section of this chapter, the fact that paradigmatic art appreciation is designer centered likewise cannot be ignored. The failure to appreciate this is one of the major ways philosophical aesthetics is misled by its disregard for the concept of appreciation and its obsession with the aesthetic.

Whatever the state of affairs in philosophical aesthetics, the point that art appreciation must focus on the artist is seemingly taken for granted by many art critics and art historians. A glance at the treatment of the point in some classic discussions of the history of art is revealing. For example, E. H. Gombrich opens his *The Story of Art* with the uncompromising claim that: "There really is no such thing as Art. There are only artists."[23] He elaborates by stressing that "what we call 'works of art' are not the results of some mysterious activity, but objects made by human beings for human beings" – objects designed such that "every one of their features is the result of a decision by the artist."[24] Indeed, Gombrich characterizes the artist as the ultimate designer, one who "must always be 'fussy' or rather fastidious to the extreme," obsessed with, as he puts it, "whether he has got it 'right'."[25] Gombrich discusses what he calls "that modest little word 'right' " at great lengths so that we may "begin to understand what artists are really after."[26]

The theme that art appreciation is designer centered is similarly evident in the introductory chapter of H. W. Janson's classic textbook, *History of Art*, aptly titled "The Artist and His Public." In discussing originality in artistic creation, Janson notes that appreciation of the ancient bronze, *Thorn Puller*

(500–400 BC), is destroyed by the knowledge that the piece is not the creation of a designing intellect, but rather an ad hoc combination of preexisting parts. He claims that with this knowledge "we no longer see it as a single, harmonious unit but as a somewhat incongruous combination."[27] When we cannot appreciate the bronze as a designed whole, as the creation of a designer, its appreciation as a work of art is not possible. Observations of the kind exemplified by Gombrich and Janson support the view that at the center of appropriate appreciation of conventional works of art is the idea that the object of appreciation is a designed object, an object created such that all its significant qualities are, as Gombrich stresses, results of decisions by a designer. This kind of appreciation, the essence of paradigmatic art appreciation, may be called design appreciation.

What are the significant features of design appreciation? The object must be appreciated as a designed object – but what does this entail? Designed objects involve three key entities: the initial design, the object embodying this design, and the individual who embodies the design in the object. With the first, the design, the central issue is what is being done or what the undertaking is or, as it is sometimes put in artistic discussions, what problem is being solved or, as it is put, rather unfortunately, in some literature in aesthetics, what the artist's intentions are. With the second, the object, what is important are its given properties, its strengths, weaknesses, limits and potentials, what can and can not be done with it, what is a use, and what an abuse, of it. With the third, the individual, what is most central are his or her abilities and skills, his or her talents, as a designer, both at the individual and the human level, and the ways these talents are used to embody the design in the object.

Design appreciation involves awareness and understanding of the three entities and their central properties. Especially important are awareness and understanding of the interplay among them and of the ways they function, severally and jointly, to achieve this interplay, for the interplay embodies the design and thereby determines the nature of the object of appreciation. In this sense design appreciation is object orientated: knowledge of the nature of the object dictates relevant acts of aspection and guides the appreciative response. A part of this response involves making judgments about the interplay, ones that assess the designed object in terms of the talents of the designer and the undertaking set by the initial design – judgments such as whether the object is a success or failure, good or bad, or, as it is often put, whether it "works" or, as Gombrich puts it, whether it is "right." Such judgments, together with the mental and physical acts of aspection required for making them, are at the core of design appreciation – and thus essential to paradigmatic art appreciation.

Appreciating art: order appreciation

Design appreciation is essential to paradigmatic art appreciation, but, as noted in the first section of this chapter, the latter must not be unthinkingly taken as the model for all other forms of appreciation. Consideration of some unconventional works of art and their nonparadigmatic yet appropriate appreciation is more helpful in understanding the appreciation of nonart and that of nature in particular. As an initial case in point, consider what is called action painting; a specific example is Jackson Pollock's *One (#31)* (1950). Its creation is described by art historian Werner Haftmann as follows:

> The canvas is placed on the ground. Casting off all intellectual control, the painter moves over it with complete spontaneity; the liquid paint dripping from his brush or from a tin with holes in it weaves the trace of his gestures into a dense filigree.[28]

Haftmann adds that a "work so produced is a direct record of the psyche" and that the process of production came to be regarded by Pollock as "pure action, an intricate trance-like choreography" the trace of which "registers the artist's inner life."[29]

What should be said about action painting concerning design, object, and artist? First, there does not seem to be an initial design that becomes embodied in the object. Haftmann says that such art is composed of and "by motor energies."[30] However, even if an embodied design is lacking, the "motor energies" yet form a pattern, which is a function of the interplay between the other two entities. Thus, Janson says of Pollock's works:

> The actual shapes visible…are largely determined by the internal dynamics of his material and his process: the viscosity of the paint, the speed and direction of its impact upon the canvas, its interaction with other layers of pigment.[31]

In this way the object and its properties achieve a greater significance than they have in more conventional art. They are no longer constrained by an attempt to make them embody a particular given design; rather they in part strongly determine a resultant pattern.

The role of the artist is more perplexing. Haftmann characterizes Pollock as moving with "complete spontaneity" and as "casting off all intellectual control." Pollock remarked in 1947: "When I am *in* my painting, I'm not aware of what I'm doing."[32] However, Pollock also noted in 1951, perhaps in response to the criticism that he, to quote Janson, "is not sufficiently in control of his medium," that: "When I am painting I have a general notion as to what I am about. I *can* control the flow of paint: there is no accident."[33] Janson's cowboy analogy is illuminating:

> Pollock does not simply "let go" and leave the rest to chance. He is himself the ultimate source of energy for these forces, and he "rides" them as a cowboy might ride a wild horse, in a frenzy of psychophysical action.[34]

Such an artist is not happily described as a *designer*, creating objects such that, as Gombrich puts it, "every one of their features is the result of a decision." Rather the artist has only "a general notion" of what he or she is about and provides "the ultimate source of energy" for the creation of the object, but not much more. The role of the artist therefore becomes similar to that of the object and its given properties: the artist is not the embodier of a design but rather only one force among others, which working together determine a pattern.

However, although in one sense the artist is assimilated to his or her materials, he or she also acquires another feature. Even if an individual is an artist, not every set of drippings, spatterings, and dribblings he or she initiates "weaves the trace of his gestures into a dense filigree." Not every pattern is an appreciable pattern. Pollock's *One*, in the Museum of Modern Art, is subtitled *#31* and as Janson says, following up his cowboy metaphor, Pollock "does not always stay in the saddle."[35] What happens when he falls out? Where are the paintings *One (#'s 1-30)*? It seems not every pattern is *selected* as an object of appreciation. Perhaps those selected are only those that seem to reveal "motor energies" or demonstrate "internal dynamics" of materials and process or actually "weave the trace" of the artist's gestures and "directly record" his or her "psyche." In short, appreciable patterns are those that reveal, or at least can be seen as revealing, an *order*, those that are or can be seen as the marks of the forces that have ordered the drips, spatters, and dribbles into whatever pattern they form. Thus, although the artist loses his or her role as designer, he or she acquires the role of selecting ordered and hence appreciable patterns.

Two additional cases, less conventional than action painting and rather more what is called anti-art, further develop the themes introduced by Pollock's work. The first involves Dada experiments with "automatic" writing and drawing and "chance" poetry and collage, especially as developed by Tristan Tzara and Hans Arp. Haftmann reports:

> Tzara would draw slips of paper with words inscribed on them from a hat, and present the resulting combination of words as a poem; Arp allowed cut-outs of free or geometric shapes to arrange themselves in a random order, then pasted them on a surface, and presented the result as a picture. In the course of such experiments, Arp also used automatic drawing, i.e., irrational, spontaneously traced forms rising from the unconscious.[36]

The key elements introduced in these experiments are spontaneity, randomness, and chance. Moreover, unlike Pollock who somewhat unconvincingly claims that "there is no accident," these artists embrace random chance as a significant part of the process. Janson claims: "The only law respected by the Dadaists was that of chance."[37] Thus, and this is especially clear concerning "automatic" works, the artist is completely reduced to a force not unlike his or her materials – a force played upon by chance. The focus is not on recording the artist's "psychophysical action" but on the spontaneous working out of the unconscious, the irrational. The artist's role is retained only insofar as there is after-the-fact alteration or, as with Pollock, selection of certain results rather than others. For example, Janson reports that Arp sometimes "cautiously adjusted" a " 'natural' configuration."[38] Moreover, some selection played a part in the overall process; after all, not every one of Arp's experiments achieved the status of *Collage with Squares Arranged According to the Laws of Chance* (1916–17), a work in the Museum of Modern Art.

The element of selection comes completely into its own only in another kind of anti-art. Although evident in the Dadaist movement, it is most perfectly exemplified in those works of Marcel Duchamp called found art. When Duchamp selected a urinal, a bottle rack, or a typewriter cover for display as a found object, selection itself became the heart of the process. Moreover, it is the fact of selection that is significant, rather than exactly what is selected, for in many instances one object does as well as another. In some surrealist experiments with found objects chance is in fact the means of selection. For example, in the following work by Salvador Dali the fact of selection is all that counts, for the object selected is left totally to chance:

> Each of the experimenters is given an alarm-watch which will go off at a time he must not know. Having this watch in his pocket, he carries on as usual and at the very instant the alarm goes off he must note where he is and what most strikingly impinges on his senses.[39]

Dadaist and surrealist art, more so than Pollock's, brings out the fact that all these works have little to do with traditional artistic and aesthetic concerns. For example, Dali claims the point of surrealist experiments is to realize "to what extent objective perception depends upon imaginative representation."[40] Haftmann reports that in general such techniques are "methods for opening the way to the store of images preserved in the unconscious. Their purpose is never to produce 'art'; all of them are conceived merely as instruments for exploring man's potentialities."[41] In a similar way, Duchamp's pieces, Dada experiments, and even Pollock's paintings seem to have little or no artistic or aesthetic basis. The urinal was not selected, in spite of what George Dickie suggests, because of "its gleaming white surface."[42] Duchamp says: "A point that I want very much to establish is that the choice of these Readymades was never dictated by aesthetic delectation. The choice was based on a reaction of *visual indifference* with a total absence of good or bad taste."[43] Likewise, the Dada experiments, as with those of surrealism, aim not at artistic or aesthetic qualities but at "new psychological discoveries" – "a release of the forces of the subconscious."[44] And Pollock's paintings, as noted, are a means of producing "a direct record of the psyche."[45]

The lack of regard for artistic and aesthetic concerns displayed by these works has ramifications for their appreciation. The works are not designed nor even selected to either solve conventional artistic problems or exemplify traditionally important aesthetic qualities such as grace or delicacy. Yet the forces of creation or selection operate such that these objects have appreciable patterns – patterns ordered by and revelatory of these forces. But the creation or selection, and hence the ordering, is typically accomplished by reference to some general ideas or beliefs having little to do with aesthetics or with art in a traditional sense. Instead the ideas or beliefs characteristically concern the way these works can reveal the nature and the order of things such as the subconscious, the unconscious, the human mind, the human condition. Thus, anything like a general criterion of aesthetic relevance is completely irrelevant, and even paradigmatic design-focused art appreciation seems quite out of place. Such appreciation is appropriate for artistically designed aesthetic objects and gets little purchase on unconventional works of art and anti-art. Some other form of appreciation is required.

The form of appreciation required may be called, in contrast to design appreciation, order appreciation, for to the extent that these works have anything comparable to a design, it is only an ordered pattern. The exact nature of such appreciation can be elaborated by further pursuing the comparison with design appreciation and the entities involved: the initial design, the object embodying the design, and the individual who embodies the design. As noted, the first is absent; there is no given design embodied in an object. However, for designed objects the initial design indicates what is being done, what the undertaking is. Thus, if this task is to be accomplished for an object with only an ordered pattern, it must be by some other means. In fact at least something similar is accomplished by the ideas and beliefs that play a role in the creation or selection of the object. Although they are typically nonartistic and nonaesthetic, they yet indicate if not what is being done, then at least what is going on in an ordered object. Thus, they help make the ordered pattern visible and intelligible somewhat as the initial design does for the designed object. Consequently, awareness and understanding of them and their interplay with the other entities is significant in order appreciation, as is the comparable knowledge of the initial design in design appreciation.

The remaining two entities, although not absent, are greatly altered. The object embodying the design no longer embodies a design and the individual who embodies the design is no longer a designer. The object is a thing shaped and molded by a combination of forces: initially, its own given properties, its strengths, weaknesses, limits and potentialities, but also the forces of random chance and of the artist acting upon it – and acting with control only barely more significant than that exercised by the object or by chance. Thus, as the third entity is only one force among many, he or she is in large part assimilated to the other forces – a part of the process and the materials rather than a master of them. But the third entity also has another role: he or she selects from what these forces, including his or her own, produce, and in selecting relies on some general ideas and beliefs in light of which the object's pattern seems visible and intelligible – can be perceived and understood as ordered by the forces that produce it. To put it another way, he or she selects by means of a general account, a story, or a theory that helps to make the object appreciable.

The additional role for the third entity is a significant factor in order appreciation. The new role is a traditional spectator role and therefore, as the artist's original role is assimilated to processes and materials, the new role is similarly assimilated to the appreciator. This is not surprising, for works such as the Dada and surrealist pieces are called experiments in part because any appreciator can conduct them; they are experiments *in appreciation*. In summary, then, in order appreciation an individual *qua appreciator* selects objects of appreciation from things around him or her. As noted, he or she does so by reference to a general nonaesthetic and nonartistic account that, by revealing the order imposed by the various forces – random and otherwise – which produce the selected objects, makes them appreciable. As in design appreciation, in order appreciation awareness and understanding of the entities involved – the order, the forces that produce it, the story that illuminates it – and the interplay among them is essential, but the focus is switched from a designed object created by an artist to an ordered object selected by an appreciator.

Appreciating nature: design appreciation

Frequently the appreciation of nature is assimilated to the appreciation of art. Such an assimilation is both a theoretical mistake and an appreciative pity. On the theoretical level, it typically involves misunderstanding not only appreciation but also one or both of art and nature. On the appreciative level, it can result in either failing to appreciate nature at all or appreciating it in an inappropriate manner – relying on the wrong information, engaging in the wrong acts of aspection, and having the wrong responses. As noted in Chapter 4, Ronald Hepburn puts this point concisely: If our "aesthetic education" instills in us "the attitudes, the tactics of approach, the expectations proper to the appreciation of art works only," we "either pay very little aesthetic heed to natural objects or else heed them in the wrong way."[46]

The theoretical mistake of assimilating nature appreciation and art appreciation, as with some other mistakes noted previously, can be traced in part to the disinterestedness tradition. However, at times in its history, especially in its beginnings, the tradition assimilated the appreciation of art to that of nature rather than the other way around. Nonetheless, whichever way it goes the assimilation is a mistake. When generated by disinterestedness, the mistake has the same roots as that of ignoring the designed nature of art, which was noted in the second section of this chapter. When the notion of disinterestedness is treated such that the object of appreciation is isolated and divorced from its interrelationships, and its appreciation is constrained by the general criterion of aesthetic relevance, the stage is set for the assimilation. According to this construal of disinterestedness, both works of art and natural objects are in their appreciation more or less severed from their natures and their histories. Thus, both kinds of objects may be appreciated in the same way: as pure aesthetic objects. The result is one form of appreciation – aesthetic appreciation – that appropriately applies to any and all kinds of things. Art appreciation and nature appreciation collapse into one.

As argued in the first section of this chapter, however, this construal of disinterestedness is a mishandling of the concept. With it, not only do all forms of appreciation collapse into one, also, as noted, all seemingly collapse into the blank, cow-like stare. Thus, the key to avoiding this path to the assimilation of art and nature appreciation is the same as that to avoiding the cow-like stare: the key is the object-orientated nature of appreciation. This understanding of appreciation, as it facilitates the recognition of the designer-centered nature of paradigmatic art appreciation, also both closes down this route to the assimilation and points toward the proper understanding of nature appreciation. However, the coast is not yet clear. Even with object-orientated appreciation and without mishandling disinterestedness, there are yet means to assimilate art and nature appreciation. Indeed, the temptation to assimilate them is seemingly so strong that one suspects that the desire for assimilation is in fact the cause and mistakes such as mishandling disinterestedness only an effect.

A second path to the assimilation of art appreciation and nature appreciation involves a tradition even more venerable than that of disinterestedness. This route recognizes the object-orientated nature of appreciation and thus follows the lead of the object. It also recognizes paradigmatic art appreciation as design focused. Thus, it takes design appreciation as basic and assimilates all other appreciation, including the aesthetic appreciation of nature, to it. All that is required to accomplish this assimilation is construing nature as the creation of a designer. And if nature is to be so construed, most cultural traditions have ample resources to support the construal. In the West many forms of theism seem adequate for the purpose and can thereby play a role in the assimilation of nature appreciation to art appreciation. As noted in Chapter 6, Nelson Potter nicely puts the point, remarking that on the theist view any account of the appreciation of art:

> ...would be directly applicable to Nature, where *God* is the artist... [since]...The theist sees the world as throughout the product of God's design and plan...For such a theist it may seem that there is *no difference* between art-appreciation and nature appreciation, since both are the products of intentional design.[47]

However, this method of assimilating art appreciation and nature appreciation also has difficulties. Potter, for example, claims it construes "God in a naive and excessively anthropomorphic fashion."[48] And consequently it fails "to realize that God is so different from man that even as we regard Nature as the product of God's handiwork," we become "aware that the model of a human artist and his intentions and his artwork is inadequate to understand our appreciation of beauty in nature."[49] He therefore opts to "bid farewell to those theists who think that conceiving of God as a divine Philip Johnson is an adequate model for the appreciation of beauty in nature."[50] Potter's conclusion is probably correct. His conclusion seemingly follows not only from his own concerns, but also from any more general lack of confidence in the theist world view. In any case it is a conclusion that has wide acceptance.

However, accepting the conclusion that nature is not usefully conceptualized as the creation of a designer leaves us with a problem concerning its appreciation – a problem that in the West has become increasingly acute since the end of the nineteenth century. In an insightful discussion of Ruskin's agonizing confrontation with it, Peter Fuller succinctly summarizes the issue: "Once the illusion that the world was the handiwork of God had been jettisoned, then the whole base of aesthetics needed to be re-examined."[51] Thus, the role of theism in the attempt to assimilate nature appreciation to art appreciation, together with what is perhaps the ultimate failure of this attempt, brings into sharp focus what may well be the central theoretical problem concerning the appreciation of nature: the problem of how to understand such appreciation given that paradigmatic art appreciation is analyzed as design appreciation, and yet nature is not to be construed as the creation of a designer.

This problem about nature appreciation is perplexing enough that it leads to radical solutions. One such solution is the nonaesthetic view of nature appreciation, introduced in Chapter 1. A version of this view is what Don Mannison calls the "Human Chauvinistic Aesthetic." Mannison contends that "Nature cannot be the object of aesthetic appreciation" by arguing that "only *human* artifacts can be objects of aesthetic judgement," for " 'artistry' is an essential component of an aesthetic judgement" – "The conceptual structure of an aesthetic judgement…includes a reference to a creator; i.e. an artist."[52] As noted in Chapter 6, a similar conclusion is drawn by Robert Elliot, who argues that "an apparently integral part of aesthetic evaluation depends on viewing the aesthetic object as an intentional object, as an artifact, as something that is shaped by the purposes and designs of its author" and that this is not possible with nature for "Nature is not a work of art."[53] As Elliot brings out, a concern of the Human Chauvinistic Aesthetic is that nature does not seem open to evaluative aesthetic judgments such as whether it is good or bad, a success or failure. This concern is legitimate, but drawing from it the conclusion that nature cannot be an object of aesthetic appreciation is without doubt an overreaction, not unlike Dickie's abandoning of the concept of the aesthetic. But even were it not, the conclusion would still be unacceptable. It offends against the truism noted in the first section of this chapter as well as in previous chapters: that aesthetic appreciation is applicable to any object of awareness whatever. Moreover, as argued in Chapter 6, it is not established by the arguments offered in its defense.[54]

What is important here, however, is seeing how the conclusion that nature cannot be an object of aesthetic appreciation stems in part from not recognizing the object-orientated nature of appreciation. As noted in the first section of this chapter, this concept, by ridding itself of anything like the general criterion of aesthetic relevance, allows for, indeed requires, different kinds of aesthetic appreciation as a function of the nature of the object of appreciation. Thus, all appreciation need not and should not be assimilated to the model of paradigmatic, design-centered art appreciation. In short, the position that nature appreciation cannot be aesthetic stems from a failure to recognize that there are legitimate alternatives to design appreciation. This fact is significant: if that position is a possible solution to the problem of how to understand nature appreciation when paradigmatic art appreciation is design focused and nature is not designed, then recognizing that there are alternatives to design appreciation makes evident a more plausible solution. Such a solution lies not in denying that the appreciation of nature can be aesthetic but rather in considering the alternatives to design appreciation. Thus, although in itself wrongheaded, the attempt to abandon the aesthetic appreciation of nature points towards a fruitful line of thought.

That nature appreciation is to be understood in light of alternatives to design appreciation is also suggested, ironically enough, by one aspect of the theist view. That view assimilates nature appreciation to art appreciation only by construing nature such that it fits the model of design appreciation. In this way the theist view brings out the fact that with object-orientated appreciation the way we construe the object of appreciation – the general account or the story we accept about it – contributes to determining the nature of its appropriate appreciation. With paradigmatic works of art this fact is obscured, for there is little dispute and therefore there are no alternative stories about them: no one doubts that they are creations of designers and typically there is considerable agreement about the details of their histories. However, when there is less clarity and agreement, as with nature, then alternative stories abound and form part of the basis for appreciation. Thus, with nature, the story accepted about it, the ideas and beliefs we have about it, are pivotal factors in its appreciation. In this way nature is similar to those works of art and anti-art for which the appropriate appreciation is, as noted in the third section of this chapter, order appreciation. Moreover, order appreciation is an alternative to design appreciation such as is required to solve the problem of how to understand the appreciation of nature without forcing it into the model of design appreciation.

Appreciating nature: order appreciation

The idea of order appreciation as a model for the appreciation of nature is worth exploration. It is suggested by more than just that our appreciation of nature, like that of the works of art and anti-art for which order appreciation is appropriate, is shaped by the stories we tell about it. It is also suggested by other similarities and relationships between such works of art and anti-art and the objects of nature. There is, for instance, the fact that the works called found objects need not be limited to urinals, bottle racks, or typewriter covers, but can be themselves natural objects, as Haftmann points out, "a root, a mussel, a stone."[55] Likewise, consider the fact that a work such as Dali's surrealist experiment is not made less significant if the "alarmwatch" goes off when the "experimenter" is, for example, in a forest rather than in his or her bathroom. If order appreciation is the form of appreciation relevant to such works, it is seemingly as relevant in the forest as it is in the bathroom.

Especially revealing of the relevance of order appreciation to the appreciation of nature are the claims of the artists who initiated these works of art and anti-art. For example, it is said such art "urges man to identify himself with nature" and is itself comparable to the objects of nature.[56] Arp, for instance, claims that: "These paintings, sculptures, objects should remain anonymous and form a part of nature's great workshop as leaves do, and clouds, animals, and men. Yes, man must once again become a part of nature."[57]

Of automatic poetry, in particular, Arp says:

> Automatic poetry comes straight out of the poet's bowels or out of any other of his organs that has accumulated reserves...He crows, swears, moans, stammers, yodels, according to his mood...Neither the Postillon of Longjumiau, nor the Alexandrian, nor grammar, nor aesthetics, nor Buddha, nor the Sixth Commandment are able to constrict him. [Nor, we might add, the general criterion of aesthetic relevance.] His poems are like nature; they stink, laugh, and rhyme like nature. Foolishness, or at least what men call foolishness, is as precious to him as a sublime piece of rhetoric. For in nature a broken twig is equal in beauty and importance to the clouds and the stars.[58]

On the assumption that order appreciation provides the correct model for the appreciation of nature, such appreciation has the following general *form*: An individual qua appreciator selects objects of appreciation from the things around him or her and focuses on the order imposed on these objects by the various forces, random and otherwise, that produce them. Moreover, the objects are selected in part by reference to a general nonaesthetic and nonartistic story that helps make them appreciable by making this order visible and intelligible. Awareness and understanding of the key entities – the order, the forces that produce it, and the account that illuminates it – and of the interplay among them dictate relevant acts of aspection and guide the appreciative response.

However, if this description indicates the general form of nature appreciation as modeled on order appreciation, it is yet only part of the complete understanding of such appreciation. As the Human Chauvinistic Aesthetic reminds us in denying that nature appreciation is aesthetic: "Nature is not a work of art." Consequently, even though modeled on order appreciation, nature appreciation differs in certain substantive ways from the appreciation of those works of art and anti-art for which order appreciation is also appropriate. Following the lead of the object, as demanded by the object-orientated nature of appreciation, helps bring out these differences.

As noted at the close of the preceding section of this chapter, the theist view that nature appreciation, like paradigmatic art appreciation, is design focused ironically suggests the alternative view that nature appreciation involves order appreciation. This is because the theist view brings out clearly the role played in nature appreciation by our story of it. In this way it also makes clear that, with order appreciation, to follow the lead of an object is to follow the lead of our story about it. And even if we, like Potter, reject the theist view of nature as designed, we are yet left with a variety of stories the lead of which we could follow. In one sense any of these stories can do the job required by order appreciation. Relevant stories derive from various religious and folk traditions, explaining nature, if not as created and designed by, as nonetheless involving in some way the actions of, for example, one all-powerful god, many lesser gods, entities such as spirits, demons, fairies, or heroes, or whatever. However, in part because such accounts lose much of their appeal without the element of design and in part because of the powerful appeal of an alternative story, these stories, with the possible exception of that of an all-powerful god, do not, I think, play a crucial role in *our* aesthetic appreciation of nature. On the contrary, as argued in the preceding three chapters of this volume, what plays this role, and has increasingly done so in the West since the seventeenth century, is the alternative account given by natural science.[59]

In light of the object-orientated nature of appreciation, the significance of the story provided by natural science to our aesthetic appreciation of nature can be further explained and justified. On the one hand, that science, natural or otherwise, is appropriate for nature appreciation gathers plausibility in that, as noted in the first section of this chapter, object-orientated appreciation is objective: it focuses on an object as what it is and as having the properties it has. And, of course, science is the paradigm of that which reveals objects for what they are and with the properties they have. Thus, science not only presents itself as the source of objective truth, it brands alternative accounts as subjective falsehood and therefore, in accord with objective appreciation, as irrelevant to aesthetic appreciation. In this way the significance of science in the aesthetic appreciation of nature is in part a legacy of the notion of disinterestedness. On the other hand, that the specific kind of science relevant to nature appreciation is natural science follows rather obviously from object-orientated appreciation. This marks one important difference between the order appreciation of nature and that of works of art and anti-art. For the latter the relevant stories, as noted in the third section of this chapter, are typically metaphysical, mystical, or psychological. However, it is no surprise that humanistic stories and the human sciences are relevant to the appreciation of artifacts, while the natural sciences are relevant to the appreciation of nature. What else should following the lead of the object indicate?

The *content* of nature appreciation can now be added to its form. Its form, as noted, centers on the entities significant to order appreciation: order, the forces that produce it, and the account that illuminates it. Its content is as follows. First, the relevant order is that typically called the natural order. Second, since there is no artist, not even one assimilated to processes and materials, the relevant forces are the forces of nature: the geological, biological, and meteorological forces that produce the natural order by shaping not only the planet but everything that inhabits it. Although these forces differ from many that shape works of art, awareness and understanding of them is vital in nature appreciation, as is knowledge of, for example, Pollock's role in appreciating his action painting or the role of chance in appreciating a Dada experiment. Third, the relevant account that makes the natural order visible and intelligible is, as noted, the story given by natural science – astronomy, physics, chemistry, biology, genetics, meteorology, geology as well as the particular explanatory theories within these sciences. For example, awareness and understanding of evolutionary theory is relevant to appreciating the natural order as revealed in flora and fauna; without such knowledge the biosphere may strike us as chaotic.

The remaining significant factor in order appreciation is the role of *selection* by an individual *qua* appreciator. And indeed in nature appreciation the appreciator must select particular objects of appreciation from, as Arp says, "nature's great workshop." However, here again there is a significant difference from the appreciation of works of art and anti-art. Although selection plays a key role in the order appreciation of these artistic endeavors, it seemingly has a somewhat less significant role in nature appreciation. Unlike the situation concerning such works, all of nature necessarily reveals the natural order. Although it may be easier to perceive and understand in some cases than in others, it is yet present in every case and can be appreciated once our awareness and understanding of the forces that produce it and the story that illuminates it are adequately developed. In this sense all nature is equally appreciable and therefore selection among all that the natural world offers is not of much ultimate importance. As Arp observes, "in nature a broken twig is equal in beauty and importance to the clouds and the stars."

If all nature "is equal in beauty and importance," this is a significant difference not simply between nature appreciation and the order appreciation of some works of art and anti-art, but between nature appreciation and art appreciation in general. However, it is in part explained by design appreciation's inappropriateness for nature. An important aspect of design appreciation is judging the object in terms of the talents of the designer and of the undertaking that is set by the initial design – assessing, as noted in the second section of this chapter, whether it is good or bad, a success or failure, or, as Gombrich puts it, whether it is "right." In short, the fact of a designer and of an initial design makes both design appreciation appropriate for conventional works of art and such judgments a key aspect of their appreciation. Likewise, the absence of these entities makes design appreciation not appropriate for nature and these kinds of judgments not an aspect of its appreciation. Natural objects are not such that their appreciation involves judging whether they are "right" in Gombrich's sense; in this sense they are all more or less equally right. Thus, to the extent that they are appreciable at all, all are more or less equally appreciable – "equal in beauty and importance." As noted in the preceding section of this chapter, that judgments of this kind do not apply to nature is part of the argument of those who support the Human Chauvinistic Aesthetic. It is now clear that what follows from this is almost the opposite of their conclusion that nature cannot be an object of aesthetic appreciation.

The appropriateness of order appreciation to nature, just as the inappropriateness of design appreciation, is also useful in further explaining the issue investigated in Chapter 6: the seeming fact that natural objects, unlike works of art, are more or less equally appreciable. In order appreciation, instead of judging a work a success or failure in light of a design and a designer, there is only appreciating an object as ordered in light of a story. And although such stories may differ, each (indeed this is much of the point of constructing them) illuminates nature as ordered – either by making its order visible and intelligible or by imposing an order on it. Thus, although such stories are in one sense nonaesthetic, in another sense they are exceedingly aesthetic. They illuminate nature as ordered and in doing so give it meaning, significance, and beauty – qualities that those who create the stories find aesthetically appealing. Thus, unlike design appreciation that focuses on aesthetic qualities that result from embodying an initial design in an object, order appreciation focuses on aesthetic qualities that result from applying an after-the-fact story to a pre-existent object. Moreover, the aesthetic qualities resulting from ordering nature in one way or another figure in the general attractiveness of one story as opposed to another. Thus, over the long run, stories develop so as to provide as much and as universal aesthetic appeal as is possible. In this way the stories that play a role in the order appreciation of nature work toward making natural objects all seem equally aesthetically appealing.[60]

Conclusion

In conclusion, some somewhat more vague and general contrasts between appreciating art and appreciating nature can also be illuminated by reference to differences between design and order appreciation. What lies behind a designed object is a designer and therefore in design appreciation part of our appreciative response is directed toward another intellect not unlike our own. The artist and design are thus something toward which we can feel the empathy and the closeness typical of our relationships with other human beings. Yet there is a distance, for it is an other, a different human being, toward which our feelings are directed. By contrast, all that lies behind an ordered object is a story, the account that illuminates the order. Thus, in order appreciation part of our appreciative response is directed towards whatever is in our story. And whether that be an all-powerful god, a folklore of demons and fairies, or a world of natural forces, it is something that is by its nature distinct from and beyond humankind, something essentially alien to us. Our appreciative response is to a mystery we will seemingly never fully comprehend. Nonetheless it is *our* story – our god, our folklore, our science – in light of which we respond; therefore there is, after all, a closeness. Perhaps not the same closeness that is typical of our relationships with other human beings, but maybe something more like that typical of our relationships with our pets – or at least our pet theories.

Given these differences between design and order appreciation, appreciating art and appreciating nature should also exemplify different kinds of ambivalences. On the one hand, in appreciating art we are aware that the work is a human creation, an artifact, and therefore ultimately open to our appreciation and to our understanding, our judgment, our mastery. However, we must also be aware that even if we master the work completely, it will yet not be our own, but someone else's creation that we master. On the other hand, in appreciating nature we are aware that the object is alien, a mystery, and therefore ultimately beyond our appreciation and beyond our understanding, our judgment, our mastery. However, we may also be aware that insofar as we manage to achieve some mastery of it, it is by means of our own beliefs, our own story, our own creation that we do so. Thus, perhaps nature is easiest to appreciate when our account of it is simplistic anthropomorphic folklore: a story of almost human gods or godlike human heroes – a story not unlike those of so-called primitive peoples who are said to feel exceptionally close to nature. In light of such folklore, nature should seem especially approachable, for it is illuminated not only by our own story but in terms of beings much like ourselves. Likewise, perhaps works of art are easiest to appreciate when they are our own creations. However, for most of us to appreciate art is to confront the quintessentially human by way of another, while to appreciate nature is to confront either an almighty god or blind natural forces by ourselves. In neither confrontation is appreciation necessarily easy, yet in each can be found aesthetic experiences of richness and power. However, it is no surprise if typically only the confrontations with nature are marked by overwhelming wonder and awe.[61]

Notes

1 Stolnitz's version of the position is in Jerome Stolnitz, *Aesthetics and Philosophy of Art Criticism, A Critical Introduction*, Boston, Houghton Mifflin, 1960. Dickie's initial and best known version of his attack is George Dickie, "The Myth of the Aesthetic Attitude," *American Philosophical Quarterly*, 1964, vol. 1, pp. 56–65.
2 Stolnitz, op. cit., p. 35.
3 Ibid., p. 42.
4 Ibid., p. 39.
5 Ibid., p. 33.
6 Ibid., pp. 37–8.

7 Ibid., p. 37.
8 Ibid., p. 37. Stolnitz is quoting Kate Hevner, "The Aesthetic Experience, A Psychological Description," *Psychological Review*, 1937, vol. 44, p. 249.
9 Dickie, op. cit., p. 61. For a fuller discussion of the issue of aesthetic relevance, see "Between Nature and Art" (in this volume, Chapter 8) and "Landscape and Literature" (in this volume, Chapter 14).
10 Stolnitz, op. cit., p. 36.
11 Ibid., p. 52.
12 Ibid., p. 53. Stolnitz puts the problem of aesthetic relevance as follows: "Is it ever 'relevant' to the aesthetic experience to have thoughts or images or bits of knowledge which are not present within the object itself? If these are ever relevant, under what conditions are they so?"
13 Ibid., p. 54.
14 George Dickie, *Art and the Aesthetic: An Institutional Analysis*, Ithaca, Cornell University Press, 1974, p. 40. In a similar fashion in "The Myth of the Aesthetic Attitude," op. cit., Dickie states that "an underlying aim of this essay is to suggest the vacuousness of the term 'aesthetic' " (p. 64).
15 Stolnitz, op. cit., p. 58. Stolnitz's brief discussion on pages 57–60 of the relevance of knowledge to appreciation is excellent.
16 I have elsewhere discussed in detail how Ziff's account of aesthetic appreciation seemingly avoids the major difficulties of both the disinterestedness tradition and alternative "conceptual" approaches. See "Critical Notice of Ziff, *Antiaesthetics, An Appreciation of the Cow with the Subtile Nose*," *Canadian Journal of Philosophy*, 1987, vol. 17, pp. 919–34.
17 Paul Ziff, "Reasons in Art Criticism," *Philosophical Turnings: Essays in Conceptual Appreciation*, Ithaca, Cornell University Press, 1966, p. 71.
18 Paul Ziff, "Anything Viewed," *Antiaesthetics, An Appreciation of the Cow with the Subtile Nose*, Dordrecht, Reidel, 1984, p. 136.
19 Ibid., p. 135. I have taken the liberty of adding commas to this quote. It is worth noting that the object-orientated nature of appreciation suggested by Ziff also becomes more explicit in Stolnitz in his later writings. In his insightful 1978 Presidential Address to The American Society for Aesthetics he characterizes disinterestedness as involving "scrupulous regard for the qualitative individuality of the object" and, just as Ziff does, speaks of the object itself as making "demands" – "demands that must be met if the thing is to be savored for what it uniquely is." (p. 411) See Jerome Stolnitz, "The Artistic and the Aesthetic 'in Interesting Times,' " *Journal of Aesthetics and Art Criticism*, 1979, vol. 37, pp. 401–13.
20 The objective residue of the notion of disinterestedness comes out clearly in Bullough's well-known discussion in which he notes that his version of disinterestedness, "psychical distance," involves in part "looking at it [the object of appreciation] 'objectively,' as it has often been called, by permitting only such reactions on our part as emphasize the 'objective' features of the experience." See Edward Bullough, "'Psychical Distance' as a Factor in Art and an Aesthetic Principle," *British Journal of Psychology*, 1912, vol. 5, reprinted in W. E. Kennick, (ed.) *Art and Philosophy: Readings in Aesthetics*, New York, St Martin's Press, 1964, p. 535.
21 Joseph Butler, *Five Sermons Preached at the Rolls Chapel* [1726], Indianapolis, Hackett Publishing Company, 1983, p. 20.
22 In the classic anti-intentionalist piece, for example, Wimsatt's and Beardsley's forbidden "external evidence" is more than simply artist's intentions; it is also knowledge, "revelations" as they say, about the artist. See W. K. Wimsatt, Jr. and Monroe C. Beardsley, "The Intentional Fallacy," *Sewanee Review*, 1946, vol. 54, pp. 468–88.

23 E.H. Gombrich, *The Story of Art*, London, Phaidon, 1950, p. 5.
24 Ibid., pp. 12–13.
25 Ibid., p. 14.
26 Ibid., p. 14.
27 H. W. Janson, *History of Art: A Survey of the Major Visual Arts from the Dawn of History to the Present Day*, Englewood Cliffs, Prentice-Hall, 1969, p. 13.
28 Werner Haftmann, *Painting in the Twentieth Century: An Analysis of the Artists and their Work*, New York, Praeger, 1965, p. 348.
29 Ibid., p. 348.
30 Ibid., p. 349.
31 Janson, op. cit., p. 540.
32 Jackson Pollock, "Three Statements, 1944–1951," in Herschel B. Chipp (ed.) *Theories of Modern Art: A Sourcebook by Artists and Critics*, Berkeley, University of California Press, 1968, p. 548.
33 Janson, op. cit., p. 540; Pollock, op. cit., p. 548.
34 Janson, op. cit., p. 540.
35 Ibid., p. 540
36 Haftmann, op. cit., p. 183.
37 Janson, op. cit., p. 534.
38 Ibid., p. 534.
39 Salvador Dali, "The Object Revealed in Surrealist Experiment," in Chipp, op. cit., p. 423.
40 Ibid., p. 423.
41 Haftmann, op. cit., p. 190.
42 George Dickie, "A Response to Cohen, The Actuality of Art," in George Dickie and R. J. Sclafani (eds) *Aesthetics: A Critical Anthology*, New York, St. Martin's Press, 1977, p. 199.
43 Marcel Duchamp, quoted in Anne d'Harnoncourt and Kynaston McShine (eds) *Marcel Duchamp*, New York, Museum of Modern Art, 1973, p. 89. The quote is given in Timothy Binkley, "Piece, Contra Aesthetics," *Journal of Aesthetics and Art Criticism*, 1977, vol. 35, p. 275. Binkley's piece is very useful concerning what he calls "Art Outside Aesthetics."
44 Haftmann, op. cit., p. 182–3
45 Ibid., p. 348.
46 R. W. Hepburn, "Aesthetic Appreciation of Nature," in H. Osborne (ed.) *Aesthetics and the Modern World*, London, Thames and Hudson, 1968, p. 53. I discuss the unfortunate consequences of appreciating nature in terms of some particular artistic models in "Appreciation and the Natural Environment," *Journal of Aesthetics and Art Criticism*, 1979, vol. 37, pp. 267–75 (reproduced in this volume, Chapter 4).
47 Nelson Potter, "Aesthetic Value in Nature and In the Arts," in Hugh Curtler (ed.) *What is Art?*, New York, Haven, 1983, pp. 142–3.
48 Ibid., p. 143.
49 Ibid., pp. 143–4.
50 Ibid., p. 144.
51 Peter Fuller, "The Geography of Mother Nature," in D. Cosgrove and S. Daniels (eds) *The Iconography of Landscape: Essays on the Symbolic Representation, Design and Use of Past Environments*, Cambridge, Cambridge University Press, 1988, p. 25.
52 Don Mannison, "A Prolegomenon to a Human Chauvinistic Aesthetic," in Don Mannison, Michael McRobbie, and Richard Routley (eds) *Environmental Philosophy*, Canberra, Australian National University, 1980, pp. 216, 212–13.
53 Robert Elliot, "Faking Nature," *Inquiry*, 1982, vol. 25, pp. 81–93. This quote is from p. 90.

54 I consider these arguments in detail in "Nature and Positive Aesthetics," *Environmental Ethics*, 1984, vol. 6, pp. 5–34 (reproduced in this volume, Chapter 6).
55 Haftmann, op. cit., p. 191.
56 Hans Arp, "Abstract Art, Concrete Art" in Chipp, op. cit., p. 391.
57 Ibid., p. 390.
58 Ibid., p. 391.
59 I argue for the importance of scientific knowledge to the aesthetic appreciation of nature in "Appreciation and the Natural Environment," op. cit. (reproduced in this volume, Chapter 4), "Nature, Aesthetic Judgment, and Objectivity," *Journal of Aesthetics and Art Criticism*, 1981, vol. 40, pp. 15–27 (reproduced in this volume, Chapter 5), "Nature and Positive Aesthetics," op. cit., (reproduced in this volume, Chapter 6), and "Saito on the Correct Aesthetic Appreciation of Nature," *Journal of Aesthetic Education*, 1986, vol. 20, pp. 85–93.
60 I develop the line of thought of this paragraph more fully in "Nature and Positive Aesthetics," op. cit. (reproduced in this volume, Chapter 6) and in "Critical Notice of Rolston, *Philosophy Gone Wild*," *Environmental Ethics*, 1986, vol. 8, pp. 163–77.
61 A segment of this chapter was presented at a symposium on "Artistic and Natural Beauty" at the American Philosophical Association Meeting, Portland, Oregon, March, 1988. I thank those present for useful comments.

Part II

LANDSCAPES, ART, AND ARCHITECTURE

8

BETWEEN NATURE AND ART

The question of aesthetic relevance

In the second part of this volume, I turn from the aesthetics of nature to a set of appreciative issues that arise, as it were, between nature and art. I address questions about the appreciation of landscapes, such as those of agriculture, which exist in the space between these two realms; I consider aesthetic problems that are generated by intimate relationships between nature and art, as in the cases of gardens and environmental works of art; and I reflect on appreciative issues that are posed by art forms, such as architecture, that fall near the boundary of art.

One general way to frame this diverse set of issues is by reference to the question introduced in Chapter 7: the question of aesthetic relevance. I thus develop this question more fully in this chapter, such that it may stand as a backdrop to the issues taken up in the next five chapters. I return to it explicitly only in the final chapter of the volume. The question itself constitutes an ideal backdrop for what follows in that it poses one of the most central and significant issues in aesthetics. It is significant from both the theoretical and the applied point of view. Its theoretical significance stems from the fact that answers to the question frequently rest upon philosophical positions of the most fundamental kind. Its significance for applied aesthetics derives from the fact that answers to the question dictate the ways and means of our actual appreciative practice. Thus the question stands as a bridge between what we think and what we do, between our aesthetics and our appreciation.

As indicated in the preceding chapter, the question is that of what is relevant to the appropriate aesthetic appreciation of any particular object of such appreciation. But since aesthetic appreciation by its nature focuses on what an object presents to the senses, the question is more precisely that of what, if anything, of that which an object does not present to the senses is relevant to its appropriate appreciation. Any appreciator confronts a potential object of appreciation with a head full of thoughts, images, associations, and bits of information. The question of aesthetic relevance is the question of what, if any, of all this "outside information" is relevant to the appreciator's appropriate appreciation.[1]

Traditional aesthetics offers a rather conservative answer to the question of aesthetic relevance: little outside of what is presented to the senses is taken to be relevant. As elaborated in Chapter 2, this lean answer rests upon two of the most fundamental philosophical positions in modern aesthetics: the doctrines of disinterestedness and formalism. The former doctrine, as noted in Chapter 7, typically involves a special attitude or state of mind in virtue of which the appreciator is disinterested in or distanced from the object of appreciation. This attitude or state thus gives a standard of aesthetic relevance that characteristically mandates a mode of appreciation in which both the appreciator and the object are isolated and separated from the world at large. Thereby, appropriate aesthetic appreciation is severed from anything that is not present in the object itself, and information external to the object is deemed irrelevant to such appreciation.[2]

The second theoretical underpinning of the conservative answer to the question of aesthetic relevance is formalism. As indicated in Chapters 2 and 3, formalism provides an essentialist analysis of art in terms of formal relationships of shapes, lines, and colors, which are typically claimed to evoke a special emotion in the appreciator. Moreover, formalism sharply distinguishes this essential formal dimension of art from everything else. Art and its appreciation is confined to a world of its own, with emotions of its own, and unrelated to the world at large. The appreciator is to bring to the object of appreciation nothing from this outside world. Appropriate aesthetic appreciation is again severed from anything that is not present in the object itself, and information external to the object is again deemed irrelevant to such appreciation.[3]

Thus supported by essentialistic formalism and the doctrine of disinterestedness, the conservative answer to the question of aesthetic relevance has severe and austere consequences for aesthetic appreciation. Thoughts, images, associations or information not presented to the senses by the object itself are condemned as irrelevant and even harmful, best purged from our minds if we are to appropriately appreciate an object for what it really is. This approach, however, has not continued unchallenged into the present, for the history of analytic aesthetics is in part a history of the assault on formalism and disinterestedness as well as on essentialism; as these philosophical positions have fallen from favor, the conservative answer to the question of aesthetic relevance has followed suit. Not only formalism and disinterestedness, but also closely related dogmas such as aestheticism, anti-intentionalism, and the autonomy of art have all been thoroughly critiqued and, in many quarters, rejected.

The upshot of this critique of the foundations of the conservative approach to the question of aesthetic relevance is that answers have become increasingly liberal. Thus, for example, in a recent article, Marcia Eaton offers the following test for aesthetic relevance: "A statement (or gesture) is aesthetically relevant if and only if it draws attention (perception, reflection) to an aesthetic property."[4] She elaborates saying that "whatever directs attention to intrinsic features aesthetically valued is aesthetically relevant" and, agreeing with Richard Wollheim, claims this may be "information of any sort whatsoever."[5] In short, if it works for the appreciator, it's relevant. Moreover, there are even more extreme answers to the question of aesthetic relevance. Following in the wake of analytic aesthetics' assault on essentialism, formalism, and disinterestedness, more recent philosophical movements such as feminist aesthetics and postmodernism so completely reject these doctrines that they leave no basis for anything but the most radical answers to the question of aesthetic relevance. Thus we now have what I call, rather casually and indifferently, a postmodern approach to aesthetic relevance, which seemingly holds that anything and everything that any appreciator happens to bring to an object is relevant to its aesthetic appreciation, for in some sense or other the object is created anew in light of that which the appreciator contributes to it.

Objects of appreciation and aesthetic necessity

Does calling into question formalism and disinterestedness inevitably lead to a postmodern approach to aesthetic relevance? Is there no stable middle ground between the extreme purism required by the former doctrines and the excessive liberalism offered by the latter approach? Must we say that either next to nothing or, alternatively, almost anything is relevant to our appropriate aesthetic appreciation of any object?

Perhaps there is a stable middle ground between these two extremes. However, establishing such an alternative requires first identifying and abandoning an assumption common not only to formalism and the doctrine of disinterestedness, but also to both their analytic and their postmodern detractors. What all these views have in common is a focus on the subject of appreciation, the appreciator, rather than on the object of appreciation. This is in part because they all, even the latest postmodern protestations, stand surprisingly firmly in an aesthetic tradition that stretches back to Kant and beyond. They retain a commitment rooted in the subjectification of aesthetic experience that was so successfully accomplished by the British empiricist aestheticians of the eighteenth century. The commitment is clear concerning disinterestedness, which is itself a special mental state of the appreciator, and formalism, which ties form to a special emotion of the appreciator. But even a refined and late-in-the-day analytic approach such as Eaton's is focused almost exclusively on achieving the right attentional state in the appreciator, however that may be accomplished. And the postmodern approach seemingly simply turns the whole business over to the appreciator and whatever state he or she happens to be in.

But why should the question of aesthetic relevance be addressed by focusing on the states of the appreciator? Isn't it rather odd to think, along with Eaton and Wollheim, for example, that whatever – "information of any sort whatsoever" – happens to put an appreciator of any particular object in a state such that he or she can appreciate that object is thereby relevant to the appreciation of the particular object in question? It follows from this view that if we could find one single bit of information, perhaps a Zen insight, or even one simple drug, perhaps a mild hallucinogenic, that would put appreciators in a state to aesthetically appreciation anything whatsoever, then that bit of information, or drug, or whatever would be aesthetically relevant to everything whatsoever. But this is surely a vacuous answer to the question of aesthetic relevance and a reductio of the attempt to answer that question primarily by reference to states of the appreciator.

A more fruitful approach to the question of aesthetic relevance is to tie its answer not to the subject of appreciation, but rather, as elaborated in the preceding chapter, to the object of appreciation. As noted in that chapter, Paul Ziff puts this point by saying that objects of appreciation "make demands." And, since objects of appreciation obviously differ, so does what is demanded for their aesthetic appreciation.[6] However, what this means is that answers to the question of aesthetic relevance must be given by reference to the nature of objects of appreciation rather than states of appreciators. Recognizing the fact that objects of appreciation "make demands" also has significant implications for how the question of aesthetic relevance itself is understood. When the question is put in terms of states of the appreciator, it is typically taken as the question of what "outside information," if any, will put the appreciator into the required state. And thus the answers tend to cluster around two extremes, either that none whatsoever is needed or that any whatsoever will do – just as long as it does the job. But when the question is put in terms of the object of appreciation, it must be understood as the question of what "outside information" is *necessary* in order to appreciate this particular object or objects of this particular kind. In short, the question of aesthetic relevance must be interpreted as a question of aesthetic necessity.[7] When the question is understood in this manner, both of the extreme answers, none whatsoever and any whatsoever, seem completely implausible. Given a particular object, why would one imagine that no information about it is necessary for its appreciation, or, even less plausibly, that any information whatsoever is?

Thus, the object-focused approach to the question of aesthetic relevance, in interpreting the question as one of what is necessary for the appropriate appreciation of the object in question, has direct implications for the traditional answers to the question. On the one hand, the turn to the object immediately takes the question of aesthetic relevance beyond the conservative answer supported by doctrines such as formalism and disinterestedness. Yet, on the other hand, concerning liberal answers such as Eaton's or that of postmodernism, the object-focused approach does not bring us even as far as the former, let alone the latter. This is because the key idea of objects making demands on appreciators involves not only requirements for appreciators, but also limitations on them: appropriate aesthetic appreciation of an object is understood as appreciation in light of what is necessary, but not more than this. There remains irrelevant as well as relevant "outside information" for the appreciation of any given object. Thus, an object-focused approach steers between the two extremes, rejecting the all or nothing dichotomy that has captivated the philosophical tradition.

Between nature and art: appreciating other things

The object-focused approach to the question of aesthetic relevance not only moves beyond the traditional philosophical answers to the question, it also has clear and important implications concerning applied aesthetics. In recognizing that an object of appreciation "makes demands" on appreciators, it establishes the necessity of "outside information" in our actual appreciative practice. This should be obvious concerning the two paradigm kinds of objects of aesthetic appreciation: art and nature. Concerning the former, it should be clear, as is assumed in the first part of this volume, that in appropriate aesthetic appreciation of works of art certain historical, cultural, and artistic information is necessary. The information concerns answers to questions such as what the work is, how it was produced, and why it was produced in the way it was. Such information may be called the "history of production" of the work. However, although these ramifications of an object-focused approach concerning art appreciation are significant, they are hardly novel. Few but the last supporters of strict formalism would deny the aesthetic relevance of information about the histories of production of works of art, and few but the advocates of the most radical postmodernism would embrace the relevance of absolutely any information whatsoever.

Concerning the other paradigm kind of object of aesthetic appreciation, nature, similar conclusions are plausible. Indeed, this is one of the major contentions of the first part of this volume, especially Chapters 4, 5, and 6. To briefly reiterate that contention: it is not unreasonable to believe that, in the aesthetic appreciation of nature, information about "history of production" is equally as aesthetically relevant as it is in the case of art. The difference is that with nature the relevant information – the story of nature's history of production – is scientific information about the natural world. Thus, as the information provided by art critics and art historians is aesthetically relevant for art, that provided by naturalists, ecologists, geologists, and natural historians is equally so for nature.[8] Moreover, as in the case of art, natural history's aesthetic relevance to the appropriate appreciation of the natural world is becoming increasing accepted.[9] And there is even awareness that some approaches to the appreciation of nature – for example, those involving excessive emphasis on ideas that are less naturalistic, such as the picturesque – may involve aesthetically irrelevant thoughts, images, and associations.

What, however, about the aesthetics of the never-never land between nature and art? What does the object-focused approach say about appropriate aesthetic appreciation of, for example, roadsides and cornfields, gardens and graffiti, churches and office buildings? To address this issue, it is useful to more fully understand the insight of the object-focused approach concerning nature and art. For each of these, this approach answers the question of aesthetic relevance by reference to history of production: the explanation of how the object of appreciation came to be as it is. For each of nature and art such explanation is the core of what is necessary for appropriate aesthetic appreciation. This is because each of at least the extremes of nature and art – pristine nature and pure art – are what they are primarily in virtue of how they came to be as they are. In other words, neither pristine nature nor pure art have, as such, a purpose or a function. They are not what they are in virtue of what they are meant or intended to accomplish. This, of course, does not mean that particular works of art and particular natural objects cannot be put to a purpose, but only that having such a purpose is not a dimension of their natures as nature and as art. Thus, for nature and art, the question of aesthetic relevance is successfully addressed by reference to history of production and perhaps even by reference to this alone.

However, once an appreciator moves the slightest distance from either pristine nature or pure art in the direction of the arena of human endeavor that stands between these two extremes, the situation changes dramatically. Any movement away from pristine nature takes one into the world of landscapes, countrysides, farmsteads, and beyond. Any movement away from pure art brings one to architecture, industrial design, commercial art, and all the varied applied arts. Of course, the fact that such endeavors constitute neither pristine nature nor pure art, and, as such, are not purposeless, in no way makes them inappropriate objects for aesthetic appreciation. They are fully fledged, fully qualified objects of aesthetic appreciation and the question of aesthetic relevance arises as seriously for them as it does for both nature and art. What this fact does indicate, however, is that at least some of such objects are, in a sense that becomes clear in the second part of this volume, somewhat more difficult and complex objects of aesthetic appreciation.[10] In part this means that the question of aesthetic relevance cannot be successfully addressed as simply as it can for nature and art.

What, then, are the ramifications concerning the question of aesthetic relevance? Things such as roadsides, cornfields, gardens, graffiti, churches, and office buildings all belong to the rich world between nature and art, to the world of landscapes, architecture, and the applied arts. All such things have a function, a purpose; and they are what they are in virtue of what they are meant or intended to accomplish. Consequently information simply about their histories of production is typically not sufficient for addressing the question of aesthetic relevance. Rather, given an object-focused approach together with the fact that these things are functional things, what is absolutely necessary for and central to their appropriate aesthetic appreciation is information about their functions. With all such things the main issue is not simply how they came to be as they are, but why they came to be as they are. The key to their natures is the purpose or the function they are meant to serve. In appropriate aesthetic appreciation of these things, the first and most important request for "outside information" is not, as it is for both pristine nature and pure art, "What is it and how did it come to be as it is?" Rather the primary question is "What does it do and why does it do it?"

Of course, even without embracing an object-focused approach, this observation about the aesthetic relevance of function should be somewhat self-evident. In fact, it has not been completely overlooked by the tradition concerning at least some things that are neither pristine nature nor pure art. For example, the idea that there is a significant relationship between function and design has a long history in the aesthetics of architecture, and thus it seems evident that information about the function of a building is relevant to its appropriate appreciation. The idea is perhaps most famously captured in the slogan, "form follows function."[11] In a similar way, in other areas of human endeavor closely allied with pure art, such as commercial art and industrial design, there is recognition of the relevance of function to good design and thus at least implicit recognition of the necessity of information about function for appropriate aesthetic appreciation. Moreover, there is increasing acknowledgement of this kind of idea in those areas involving design on a larger scale, such as landscape architecture, city planning, and environmental design.[12]

The aesthetic relevance of information about function, however, is less clearly and fully appreciated concerning those things not closely allied to pure art, but more akin to pristine nature. Consider, for example, agricultural landscapes and rural countrysides. Although they are frequently almost completely formed in virtue of human functions and purposes, it is not always obvious that they are so formed – at least not nearly as obvious as in the case of the products of architecture and industrial design. Thus, the relevance of information about function to aesthetic appreciation is easily overlooked. Yet, to appropriately appreciate a landscape, we must know the different uses to which the land has been put and which have thus shaped its look. Some of these uses, such as the practical uses of agriculture and mining, determine the look of the land in rather evident ways.[13] But others, such the symbolic, mythical, and religious uses to which human beings put the land, contribute to the formation of landscapes in a much more subtle fashion.[14] Nonetheless, information about all such functions, whether obvious or subtle in their effects, is essential to appropriate aesthetic appreciation.

In summary: my suggestion is that in general an object-focused approach to the question of aesthetic relevance is required, for we will be guided to appropriate aesthetic appreciation only by taking into account the real nature of the object of appreciation. Moreover, concerning those things that are between nature and art, what is needed for appropriate appreciation is the recognition of the necessity of information not only about their natures but also about their functions. The chapters in the second part of this volume pursue this suggestion.

Notes

1 As indicated in Chapter 7, Jerome Stolnitz's classic formulation of the question is: "Is it ever 'relevant' to aesthetic experience to have thoughts or images or bits of knowledge which are not present within the object itself? If these are ever relevant, under what conditions are they so?" See Stolnitz, *Aesthetics and Philosophy of Art Criticism: A Critical Introduction*, Boston, Houghton Mifflin, 1960, p. 53. For a fuller discussion of Stolnitz's position and the question of aesthetic relevance, see my chapter on "Appreciating Art and Appreciating Nature," in S. Kemal and I. Gaskell (eds) *Landscape, Natural Beauty, and the Arts*, Cambridge, Cambridge University Press, 1993, pp. 199–227 (reproduced in this volume, Chapter 7).

2 I also discuss disinterestedness and aesthetic appreciation in "Understanding and Aesthetic Experience" (in this volume, Chapter 2).

3 For a fuller examination of formalism and aesthetic appreciation, see "Understanding and Aesthetic Experience" (in this volume, Chapter 2), and "Formal Qualities in the Natural Environment," *Journal of Aesthetic Education*, 1979, vol. 13, pp. 99–114 (reproduced in this volume, Chapter 3). I also discuss these issues in "On the Possibility of Quantifying Scenic Beauty," *Landscape Planning*, 1977, vol. 4, pp. 131–72.

4 Marcia Muelder Eaton, "Where's the Spear? The Question of Aesthetic Relevance," *British Journal of Aesthetics*, 1992, vol. 32, pp. 1–12. The quote is from page 4. Eaton's analysis of the aesthetic is developed in her *Aesthetics and the Good Life*, Cranbury, Associated University Press, 1989.

5 Eaton, "Where's the Spear?," op. cit., p. 5. Richard Wollheim's position is developed in *Painting as an Art*, Princeton, Princeton University Press, 1987.

6 Paul Ziff, *Antiaesthetics: An Appreciation of the Cow with the Subtile Nose*, Dordrecht, Reidel, 1984, p. 135. I elaborate Ziff's account of aesthetic appreciation in "Critical Notice of Ziff, *Antiaesthetics: An Appreciation of the Cow with the Subtile Nose*," *Canadian Journal of Philosophy*, 1987, vol. 17, pp. 919–33. I more fully discuss his contribution to an object-focused notion of appreciation in "Appreciating Art and Appreciating Nature," in Kemal and Gaskell, op. cit., pp. 199–227 (reproduced in this volume, Chapter 7) and "Nature, Aesthetic Appreciation, and Knowledge," *Journal of Aesthetics and Art Criticism*, 1995, vol. 53, pp. 393–400.

7 Eaton's excellent discussion of "aesthetic necessity" is relevant here. See Eaton, "Where's the Spear?," op. cit., pp. 6–8.

8 I develop the view that scientific information is relevant to aesthetic appreciation of nature in several other papers, reproduced throughout Part I of this volume, especially in "Appreciation and the Natural Environment," *Journal of Aesthetics and Art Criticism*, 1979, vol. 37, pp. 267–76 (reproduced in this volume, Chapter 4), "Nature, Aesthetic Judgment, and Objectivity," *Journal of Aesthetics and Art Criticism*, 1981, vol. 40, pp. 15–27 (reproduced in this volume, Chapter 5) and "Nature and Positive Aesthetics," *Environmental Ethics*, 1984, vol. 6. pp. 5–34 (reproduced in this volume, Chapter 6).

9 Of particular interest in this regard are H. Rolston, "Does Aesthetic Appreciation of Nature Need to be Science Based?," *British Journal of Aesthetics*, 1995, vol. 35, pp. 374–86 and M. Eaton, "Fact and Fiction in the Aesthetic Appreciation of Nature," in A. Berleant and A. Carlson (eds) *The Journal of Aesthetics and Art Criticism. Special Issue: Environmental Aesthetics*, 1998, vol. 56, pp. 149–56.

10 I discuss this issue in "Is Environmental Art an Aesthetic Affront to Nature?," *Canadian Journal of Philosophy*, 1986, vol. 16, pp. 635–50 (reproduced in this volume, Chapter 10), "Interactions between Art and Nature: Environmental Art," in P. McCormick (ed.) *The Reasons of Art: L'Art a ses raisons*, Ottawa, University of Ottawa Press, 1985, pp. 222–31 and "On the Aesthetic Appreciation of Japanese Gardens," *The British Journal of Aesthetics*, 1997, vol. 37, pp. 47–56 (reproduced in this volume, Chapter 11).

11 The slogan is usually attributed to Louis Sullivan. I discuss Sullivan and the importance of function in architecture in "Existence, Location, and Function: The Appreciation of Architecture," in M. Mitias (ed.) *Philosophy and Architecture*, Amsterdam, Rodopi, 1994, pp. 141–64 (reproduced in this volume, Chapter 13).

12 I discuss related matters in "Reconsidering the Aesthetics of Architecture," *Journal of Aesthetic Education*, 1986, vol. 20, pp. 21–7.

13 I address the role of information about function in the appreciation of such landscapes in "On Appreciating Agricultural Landscapes," *Journal of Aesthetics and Art Criticism*, 1985, vol. 43, pp. 301–12 (reproduced in this volume, Chapter 12).

14 I pursue these ideas in "Landscape and Literature" (in this volume, Chapter 14).

9

ENVIRONMENTAL AESTHETICS AND THE DILEMMA OF AESTHETIC EDUCATION

The eyesore argument

In this chapter I consider the great expanse of our environment that stands between pristine nature and pure art by addressing a particular environmental movement and one specific argument that is given in support of this movement. The movement is the "clean up the environment" movement and the argument is what I call the "eyesore argument."

The clean up the environment movement is, of course, that movement urging us to clean up our environment of the junk, litter, and debris that presently clutter it. It is clearly an important movement with far-reaching consequences. It expresses itself in numerous ways in our society; for example, in anti-litter campaigns in the media, in the call to recycle common litter articles such as cans, bottles, and paper, and in social pressure and legislation concerning billboards, junk yards, and mining practices. However, perhaps one of the best publicized developments of the general movement is the call to "beautify" our roads and highways.[1] For this reason I utilize the campaign to clean up our roadways as my main example. However, the issues under discussion are equally relevant to many of the other directions this environmental movement has taken and can take.

The eyesore argument is one of a number of different arguments frequently offered in support of the clean up the environment movement. It does not, as some other arguments do, take note of contentions such as, for example, that junk, litter, and debris have a negative ecological effect or that littering is wasteful. Rather it appears to be an argument that makes an explicit appeal to aesthetic considerations. In its popular form it is the contention that we should clean the environment of junk, litter, and debris because such materials are "eyesores." I take it that the substance of this argument is simply that roadside clutter, for example, is unsightly, an eyesore; consequently, when scattered about the environment, it distracts from the aesthetically pleasing nature of that environment. Since an aesthetically pleasing environment is to be preferred to one that is not, we should clean up the environment. Understood in this way, the eyesore argument has two basic premises: first, that roadside clutter, for example, is not aesthetically pleasing and, second, that an aesthetically pleasing environment is to be preferred to one that is not. These two premises are my main concern.

The above argument is generally taken to be persuasive. However, I think that the grounds of its strength have not been made explicit. In the remainder of this chapter I attempt to bring out the argument's full strength by discussing a line of attack on the argument and some alternatives for countering this attack.

The dilemma of aesthetic education

The attack on the eyesore argument can be elaborated by considering what Monroe Beardsley calls the "dilemma of aesthetic education."[2] As I understand this dilemma, it can be developed in at least two different ways: either by reference to traditional aesthetic attitude theories or by reference to what is called "camp sensibility." Beardsley relies on the latter mode of development and I follow him in this. The idea of camp sensibility is popularized by art critic Susan Sontag. In "Notes on 'Camp' " Sontag states that "camp is a certain mode of aestheticism…It is *one* way of seeing the world as an aesthetic experience" that "has the power to transform experience."[3] Moreover, this transformation of experience is often of our experiences of objects that, although not commonly experienced as aesthetically pleasing, can be so experienced with the aid of camp. In other words, objects that are typically seen as unsightly or as in bad taste or even as simply aesthetically uninteresting can by means of camp sensibility become objects of aesthetic enjoyment. Sontag gives a number of examples that she takes to be of this kind. They include such things as the "concoctions of Tin Pan Alley and Liverpool," "turn-of-the-century picture postcards," "old Flash Gordon comics," and "kitsch art" in general.[4] It is the transformation of our experience of such objects that generates Beardsley's dilemma. The dilemma is that we are divided between two conflicting ways of dealing with something that we initially do not aesthetically enjoy: one is to change the world such that the object of aesthetic displeasure is eliminated; the other is to educate people to change *their* aesthetic sensibilities such that the object, although itself unchanged, can be experienced as aesthetically pleasing. In short, camp's transformation of our experience of aesthetically displeasing objects yields an alternative to ridding the world of such objects.[5]

The dilemma of aesthetic education as generated by camp thus meets the eyesore argument head on. It recognizes the first basic premise – that roadside clutter, for example, is not aesthetically pleasing – by admitting that we initially find litter, junk yards, strip mines, and so forth unsightly. But it counters the argument by presenting an alternative to the conclusion that the environmentalist wishes to draw – that is, that we should clean up the environment. Instead, it suggests, why not develop our camp sensibility such that roadside clutter and the like become aesthetically pleasing? Transforming our experience of the offending objects can solve the problem as effectively as removing the objects. The line of thought is suggested by a cartoon that Beardsley describes. It shows "the proprietor of a junk yard named 'Sam's Salvage' standing by a huge pile of junked cars, and saying to two other men: 'Whattya mean it's an ugly eyesore? If I'd paid Picasso to pile it up, you'd call it a work of art.' "[6] Seeing something as a work of art can do the same job as camp sensibility.

The proposal that we solve the problem of our unsightly environment simply by changing our sensibilities is likely to be condemned as too ridiculous to consider. However, some things can be said in its favor. First, in recent years camp has become a rather well-respected form of sensibility for certain art genres. Sontag, for example, points out that a full analysis of art nouveau cannot ignore that aspect of it that lends itself to camp. Moreover, although Sontag does admit that "not everything can be seen as camp," camp sensibility seems particularly appropriate to many of those objects that constitute roadside clutter. This is suggested by the fact that camp has developed hand in hand with certain art movements, some of which often imitate such things as billboards and tin cans, and others of which occasionally utilize junk and trash as a medium. Second, there are some practical points in favor of this proposal. For example, developing camp sensibility toward our environment might be easier and would certainly be more economical than attempting to clean up the environment. Perhaps we as a society cannot afford to select the latter alternative? And given that there are some segments of our environment that virtually cannot be salvaged, adopting camp sensibility for them is certainly more reasonable than forcing ourselves to live with what we experience as an eyesore. In short, the use of camp for our so-called unsightly environment seems respectable, appropriate, practical, economical, and in some cases our only hope! With all this in favor of camp sensibility, the eyesore argument, by contrast, does not appear to give much support for cleaning up the environment.

Nonetheless, in spite of almost anything that can be said for the camp sensibility proposal, I think it is clear that most people, and especially environmentalists, would not find it a satisfactory answer to the environmental problems under discussion here. They simply would not see it as an acceptable response to the eyesore argument. We should not conclude from this that environmentalists and like-minded people are narrow-mindedly wanting to solve our environmental problems in their own way, without regard for somewhat creative new proposals. Rather we should attempt to take a more serious look at these issues and especially at the eyesore argument itself.

The natural

As noted previously, the eyesore argument has two basic premises. The first – that roadside clutter is not aesthetically pleasing – is the focus for the counter to the argument posed by the dilemma of aesthetic education. We must ultimately discuss this premise further, but first it is helpful to consider the

people actually have in mind when they present what sounds like the eyesore argument. First, we can note that this new argument is not affected by the dilemma of aesthetic education. Second, however, it unfortunately has other problems, many of which stem from the vagueness of the term "natural." For example, it is not obvious that trees and shrubs are the *natural* environment for the sides of roadways and the outskirts of cities; in some senses of "natural" perhaps junk yards, litter dumps, and billboards are much more natural. The point is that arguments that appeal to the natural concerning environmental problems necessarily presume answers to questions such as "What kinds of things are natural?" and "What is man's natural environment?" These are important philosophical questions that must be considered before any argument from naturalness can be evaluated. However, to discuss them here would take us too far from our aesthetic-centered concerns.[9]

The aesthetically pleasing

The digression of the preceding section shows that the eyesore argument cannot easily be salvaged through elaborating its second premise by reference to naturalness. The result is either a new set of problems or a quite different argument – an argument that is also problematic. Consequently, we must return to the first premise – that roadside clutter is not aesthetically pleasing – and examine a key concept of both this and the second premise: the concept of being aesthetically pleasing.

Up to this point we have relied on a rather pre-analytic, intuitive notion of being aesthetically pleasing; one that is, I suspect, quite similar to that used in the popular form of the eyesore argument. However, to shed light on the problems posed by the dilemma of aesthetic education, we need to refine this notion in certain ways. The tradition of philosophical aesthetics provides a distinction and some accompanying concepts that are helpful. The distinction is related to some of the issues discussed in Chapters 2 and 3; here we may understand it as a distinction between two senses of being aesthetically pleasing. In this form the distinction is well-illustrated in the work of D. W. Prall and, following him, John Hospers.[10] Hospers describes it as a distinction between the "thin sense" and the "thick sense" of "aesthetic," labels that serve for our purposes. The thin sense is relevant when we aesthetically enjoy an object primarily in virtue of the physical appearance of the object, including not only its surface physical properties, but also its formal properties having to do with line, shape, and color. The thick sense, on the other hand, involves not merely the physical appearance of the object, but also certain qualities and values that the object expresses or conveys to the viewer. Prall calls this the "expressive beauty" of the object, while Hospers speaks of objects expressing "life values."

The distinction can be elaborated in terms of examples. Consider an older house. We often find such houses aesthetically pleasing because of, for example, the design of the windows or the color of the woodwork, but this is only part of the matter. In many cases we also enjoy such houses aesthetically because they give the general impression of a less hectic, more genteel way of life or show more signs of care and craftsmanship than do many newer houses. In like manner, much music is aesthetically pleasing not only because of its intricate pattern of sounds, but also because of the melancholy or the sadness or the joy it expresses; some sports cars are aesthetically pleasing not only in virtue of their lines and colors, but also in virtue of expressing speed and workmanship. Hospers gives some similar examples from the natural environment:

> When we contemplate a starry night or a mountain lake we see it not merely as an arrangement of pleasing colors, shapes, and volumes, but as expressive of many things in life, drenched with the fused association of many scenes and emotions from memory and experience.[11]

These comments and examples adequately clarify, for our purposes here, the distinction between the thin and the thick sense of being aesthetically pleasing. However, before applying this distinction, some additional remarks about the thick sense are in order. Concerning this I speak of objects expressing "life values" or having "expressive qualities." These terms refer to a fairly wide range of human values, emotions, and attitudes that are associated with objects in such a way that it is appropriate to say that an object *expresses* these values, emotions, and attitudes. The relevant concept of "expression" is of the kind initially clarified by Santayana.[12] Thus for an object to express a quality or life value, the latter must not simply be suggested by it. Rather the quality must be associated with the object in such a way that it is felt or perceived to be a quality of the object itself; that is, what Santayana meant by saying that the object must seem to "embody" that which it expresses. Clarified in this way, expression is not typically due to the unique associations resulting from an individual's own personal history. Rather what is involved are the more general and deep-seated associations that are characteristically held in common by a community of individuals and by and large derived from what is perceived within that community of individuals to be the nature and function of the expressive object. Thus, the life values an object expresses are often the ones reflecting the values, emotions, and attitudes of the individuals who are responsible for its nature and function.

Life values and the eyesore argument

Given the distinction between the thin and the thick sense and accompanying concepts such as "expression" and "life values," we can now sort out the issue between the eyesore argument and camp sensibility, and perhaps reveal the full strength of the eyesore argument. If we take this argument in terms of the thin sense, the argument claims that the physical appearance and form of the environment is more aesthetically pleasing than the physical appearance and form of billboards, junk yards, roadside litter, and the like. This claim seems initially plausible, and consequently the argument is usually not developed further. In short, the argument appears cogent when developed in terms of the thin sense and is therefore not developed in terms of the thick sense. However, a dilemma is posed by camp sensibility just because camp is claimed to make the physical appearance and form of roadside clutter as aesthetically pleasing as is the environment. And if the thin sense is all that is considered, this claim seems plausible. Thus we have no cogent argument for cleaning up the environment.

However, as may be suggested by the consideration of related matters in Chapters 2 and 3, to consider this issue in terms of the thin sense alone is to do justice neither to the eyesore argument nor to camp sensibility. Whether the proponents of the eyesore argument realize it or not, roadside clutter is unsightly because of much more than its physical appearance. I think proponents of the argument tend to consider it in terms of the thin sense alone because they assume it is largely the physical appearance of the environment that makes it aesthetically pleasing. The life values expressed by the environment are less obvious and more difficult to determine. Yet the life values expressed by roadside clutter are at least as important as, if not more important than, its physical appearance in making such clutter an eyesore. In fact, I suggest that it is unsightly primarily because of these expressive qualities. By and large, its expressive qualities are qualities associated with things such as waste, disregard, carelessness, and exploitation. It is by taking note of this fact that the eyesore argument becomes a forceful argument.

In a somewhat similar manner, it is equally incorrect to consider camp sensibility only in terms of the thin sense. In the way in which it helps us to aesthetically enjoy the physical appearance and formal qualities of art forms such as art nouveau, camp can also help us to enjoy aesthetically these same aspects in billboards, beer cans, and junked cars. This is because, as Sontag says, camp emphasizes "texture, sensuous surface, and style at the expense of content."[13] Yet this is still only part of the matter. Camp sensibility is also relevant to the thick sense of an object's being aesthetically pleasing, in that camp, through its detachment and emphasis on style, often makes it possible for us to become more aware of those expressive qualities of objects that constitute the thick sense. In doing this, however, camp does not change these expressive qualities. Sontag rightly points out that "objects, being objects, don't change when they are singled out by the Camp vision."[14] What camp does is make us aware of the expressive qualities that are present and invites us to enjoy them, if we can. "Camp taste is, above all, a mode of enjoyment, of appreciation, not judgment."[15]

I noted above that the life values expressed by roadside clutter such as junk yards, strip mines, and discarded litter seem to me to be qualities associated with things such as waste, disregard, carelessness, and exploitation. If this is the case, camp sensibility cannot alter this fact; it can only make us more aware of it and ask us to enjoy aesthetically the expression of such qualities. Perhaps with camp we can find aesthetically pleasing the expression of such qualities and consequently aesthetically enjoy roadside clutter. In fact, however, I rather doubt if many of us can. A strip mine that is expressive of exploitation or roadside litter that is expressive of carelessness and disregard are at least as difficult to find aesthetically pleasing as are a sports car that is expressive of shoddy workmanship or a cheap novel that is expressive of the desire for the fast buck. And it is often due to the fact that cheap and shoddy versions of objects such as cars and houses, novels and musical compositions are expressive of such life values that we do not and cannot aesthetically enjoy them in the thick sense. If the case is similar concerning roadside clutter, then camp cannot succeed in making it aesthetically pleasing in the thick sense of that concept. The result is that with the thick sense the dilemma of aesthetic education cannot arise. This point is important, for it not only reinstates the eyesore argument as a strong and effective argument, it also indicates the full extent to which roadside and other environment clutter is truly an "eyesore."

The preceding brings out the difficulty of aesthetically enjoying in the thick sense objects that express certain life values. That we often are not able to enjoy aesthetically in the thick sense such objects is a fairly uncontroversial empirical claim. However, there is a second and perhaps more controversial empirical claim that, if accepted, makes the above line of thought even stronger. This is the claim that when we are actually unable to find an object aesthetically pleasing in the thick sense because of the (negative) nature of its expressive qualities, this often makes aesthetic enjoyment of this object in *the thin sense* psychologically difficult, if not impossible. If this claim is true, then in some instances any aesthetic enjoyment of an object will be impossible because of the nature of that object's expressive qualities. In light of this, the fact that camp sensibility often makes us *more* aware of an object's expressive qualities becomes quite significant. It means that the utilization of camp for certain objects will, in virtue of making us more aware of their expressive qualities, make any aesthetic enjoyment of these objects impossible. Moreover, since camp, by definition, is a mode of aesthetic enjoyment, the successful and sustained adoption of camp sensibility toward such objects becomes impossible. In short, some objects, because of their expressive qualities, cannot be aesthetically enjoyed by adopting camp sensibility. To attempt to do so is self-refuting. Consequently, if we grant the truth of the above-mentioned empirical claim concerning certain objects, the conclusion is not only that camp cannot generate the dilemma of aesthetic education in *either* the thin or the thick sense for such objects, but also that these objects are among those which Sontag admits cannot be "seen as camp." This conclusion gives a strong version of the eyesore argument in the thin as well as the thick sense by means of showing that, in the final analysis, we need not take seriously the challenge of camp sensibility.

I believe the foregoing line of thought yields a strengthened and yet essentially aesthetic version of the eyesore argument. However, in doing so, it depends on two claims that I take to be empirical. Since I do not wish the issue to turn completely on either of these empirical claims, I conclude this section by mentioning a second, perhaps in the long run more promising, direction for the eyesore argument to take. This direction parallels that of the moral argument concerning the aesthetic appreciation of nature that is suggested at the end of Chapter 5.

This development of the argument turns on what seems to me a conceptual point: that our aesthetic enjoyment of an object counts toward our wishing to experience that object and thus against our wishing to eliminate it. Consequently, if we find roadside clutter aesthetic pleasing, our desire to experience it is somewhat heightened and our desire to eliminate it somewhat lessened. Now suppose for a moment that with the help of camp sensibility or whatever, we do find roadside clutter aesthetically pleasing in the thick sense. This would mean that we aesthetically enjoy the expression of certain life values. Moreover, given the concept of expression discussed above, many of these life values are expressed by roadside clutter in virtue of the human values and attitudes that are in part responsible for roadside clutter. However, in light of the above-mentioned conceptual point, our aesthetic enjoyment of the expression of these life values involves at least tacitly condoning these human values and attitudes in virtue of which roadside clutter expresses these life values. In general, our not wishing to eliminate and in fact wishing to experience an effect requires at least condoning, if not actually approving, the cause. But the problem is that many of these human values and attitudes are of a kind that we find morally unacceptable, and condoning or approving the morally unacceptable is itself morally unacceptable. Thus when we find it possible to enjoy roadside clutter aesthetically, we may not find it morally acceptable to do so. We may in the last analysis be forced by our moral values to clean up the environment. I believe that this suggests a fruitful way in which the eyesore argument may be developed. Construed in this way, it is not simply an aesthetic argument, but a moral–aesthetic argument that relies on a certain combination of our moral values and our aesthetic sensibilities. I also think that in construing it in this manner we may be coming closer to the popular form of the argument, for the dialogue between environmentalists and their opponents is often marked by moral indignation and outrage, even when the issues seem to be argued purely in aesthetic terms.

Conclusion

In the preceding section of this chapter, I have outlined some ways in which the eyesore argument may be seen to have the force we initially think it has. In doing so I have relied on the assumption that roadside clutter expresses certain objectionable life values, those associated with things such as waste, disregard, carelessness, and exploitation. I conclude this chapter with some further discussion of this assumption in the hope of supporting its plausibility. Let us begin by reference to some related examples. Consider the life-size, plastic "trees" previously mentioned. I admit that in one sense such "trees" may be as aesthetically pleasing as the real thing. This is in the thin sense. If these "trees" are good replicas, they will have a physical appearance and form very similar to real trees and consequently in the thin sense be equally aesthetically pleasing. In this case, as in the roadside clutter case, however, it is expressive qualities that are important. I think that we would find plastic "trees" aesthetically unacceptable mainly because of the life values they would express. And although it is difficult to describe exactly what they would express, I suggest it would be something like a combination of resignation and ingenuity. Such an expressive quality is not as objectionable as those of roadside clutter, but it is still rather disconcerting, disconcerting enough to make such "trees" difficult to enjoy aesthetically (and yet perhaps not disconcerting enough to prevent them from being paradigm objects for camp sensibility).[16]

A similarly revealing example is that of "junk art." By this phrase, I mean art that is constructed from at least some of the possible contents of roadside clutter. Examples are sculptures such as Picasso's *Bull's Head* (1943), a "bull's head" constructed of bicycle parts, and John Chamberlain's *Essex* (1960), made with automobile parts and scrap metal. Whether or not such art is aesthetically pleasing depends to a great extent on its expressive qualities and on the expressive qualities of the materials from which it is constructed. On the one hand, when we find such art distasteful, this is often because the materials have kept their original expressive qualities. The artist has not reworked the materials so as to prevent the art work from expressing rather questionable life values. The obvious examples of this kind of case are, I expect, some of what is called "found art." Perhaps this is why some people find distasteful works such as, for example, Duchamp's *Fountain* (1917) – the famous urinal placed in an art show. On the other hand, we often find junk art aesthetically pleasing (and morally satisfying). Such is the case, I believe, with *Bull's Head*. This is, I suggest, because here the artist has in effect "recycled" the materials he has utilized and in doing so changed their expressive qualities. Discarded objects that would otherwise express waste and disregard are reworked such that they now, as a work of art, express utilization, concern, and sensitivity. With this in mind, I think we can shed light on the cartoon the description of which I quoted from Beardsley above. The point is that had the pile of junked cars been piled by Picasso, it may well have expressed different life values; it may have expressed the qualities of junk "recycled" as art, rather than the life values typically expressed by objects that have been used, abused, and discarded by a waste-oriented society.

The last point I mention can be put as an objection to my suggestion that strip mines, junk yards, discarded beer cans, and the like express qualities associated with things such as waste, disregard, carelessness, exploitation. An objector might contend that such objects do not express these life values, but actually others that are aesthetically acceptable (and morally appealing), such as, for example, hard work, determination, vision. And that if this is the case, we can aesthetically enjoy these objects in both the thin and the thick sense of that concept. Therefore, the moral–aesthetic version of the eyesore argument does not apply. In reply I offer two brief comments. First, there is no inconsistency in these objects expressing very different, seemingly opposed life values. Perhaps many express both disregard and determination. But if this is the case, the objection has no force, for the expression of acceptable life values does not cancel out the expression of objectionable ones. Second, I am not sure who can say with certainty what life values these objects express. Although I suspect that the art critic and the social critic can help, in the last analysis it is, I expect, up to us as a community of individuals. For example, it seems to me that while small family farms along the road may express determination, discarded car bodies do not, and while the skyline of a city may express vision, a strip mine does not; but these may be eccentric opinions. Thus perhaps the first step is for each of us to keep the issue in mind when we next take a look at our environment.

Notes

1 There are attempts to approach the goals of this movement in a scientific or at least quasi-scientific manner. For example, concerning roads and highways in particular, see H. Burke, C. Lewis, and H. Orr, "A Method for Classifying Scenery from a Roadway," *Park Practice Guideline*, 1968, vol. 3, pp. 125–41. For a critique of this general kind of approach, see my "On the Possibility of Quantifying Scenic Beauty," *Landscape Planning*, 1977, vol. 4, pp. 131–72.

2 Beardsley introduces this "dilemma" in Monroe Beardsley, "The Aesthetic Point of View," *Metaphilosophy*, 1970, vol. 1, pp. 39–58. He does not suggest a resolution.

3 Susan Sontag, "Notes on 'Camp'," *Against Interpretation*, New York, Dell, 1969, p. 279.

4 Ibid., pp. 279–80.

5 I noted that the dilemma can be developed by reference to traditional aesthetic attitude theories. This is because most attitude theorists hold that, as it is put by, for example, Jerome Stolnitz, "the aesthetic attitude can be adopted toward 'any object of awareness whatsoever.' " See Jerome Stolnitz, *Aesthetics and the Philosophy of Art Criticism*, New York, Houghton Mifflin, 1960, pp. 40–2. If so, the aesthetic attitude will generate the dilemma.

6 Beardsley, op. cit., p. 55. The cartoon is by David Gerard.

7 I understand that Jefferson Boulevard in Los Angeles has (or had or was at one time planned to have) plastic trees. I do not know if this is true, but it is suggested by Martin H. Krieger in "What's Wrong with Plastic Trees?", *Science*, 1973, vol. 179, pp. 446–55.

8 See, for example, W. Rutherford and E. L. Shafer, "Selection Cuts Increase Natural Beauty in Two Adirondack Forest Stands," *Journal of Forestry*, 1969, vol. 67, pp.

415–19. However, such evidence may well be tainted by an overly formalistic approach to the environment. For a discussion of the relevant issues, see my "Formal Qualities in the Natural Environment," *Journal of Aesthetic Education*, 1979, vol. 13, pp. 99–114 (reproduced in this volume, Chapter 3) and my "On the Possibility of Quantifying Scenic Beauty," op. cit.

9 It is important to note that the whole question of a correlation between naturalness and being aesthetically pleasing is much more complex than suggested in this short digression. I address some relevant issues in "Nature and Positive Aesthetics," *Environmental Ethics*, 1984, vol. 6. pp. 5–34 (reproduced in this volume, Chapter 6).

10 D. W. Prall, *Aesthetic Judgment*, New York, Thomas Y. Crowell Company, 1929, pp. 178–227 and John Hospers, *Meaning and Truth in the Arts*, Chapel Hill, University of North Carolina Press, 1946, pp. 11–15.

11 Ibid., pp. 12–13.

12 George Santayana, *The Sense of Beauty* [1896], New York, Collier Books, 1961, pp. 137ff. I utilize an analysis of "expression" similar to that of Santayana for I take it that such an analysis is, on the one hand, quite accessible, and on the other, adequate for present purposes.

13 Sontag, op. cit., p. 280.

14 Ibid., p. 284.

15 Ibid., p. 293.

16 I think the life values expressed by such things as plastic trees are significantly different from those expressed by roadside clutter for a number of reasons. One that may be particularly important, however, is that while the latter is simply another dimension of our problem, the former is a response to that problem, albeit a somewhat feeble and misguided response.

10

IS ENVIRONMENTAL ART AN AESTHETIC AFFRONT TO NATURE?

Environmental works of art

In this chapter I consider one aesthetic issue that arises between nature and art in the sense of arising from an intimate relationship between the two. There are, of course, many such relationships between nature and art in the history of humankind's domination of the earth. However, to investigate this aesthetic issue I here take as my example a quite recent and essentially artistic phenomenon – what is typically called environmental art. Within this category I focus mainly on works such as the earthworks and earthmarks of artists such as Robert Smithson, Michael Heizer, and Dennis Oppenheim and certain structures on the land such as those of Robert Morris, Michael Singer, and Christo. Some paradigm cases are Smithson's *Spiral Jetty* (1970), Heizer's *Double Negative* (1969–70), Singer's *Lily Pond Ritual Series* (1975), and Christo's *Running Fence* (1972–76).

Environmental works of art share a common feature that both distinguishes them from traditional art and makes them examples of the most intimate of relationships between nature and art. This is that all such works of art are in or on the land in such a way that a part of nature constitutes a part of the relevant aesthetic object. In other words, not only is the site of an environmental work an environmental site, but the site itself is an aspect of the work. Art critic Elizabeth Baker, for example, speaks of the site as "a part of the content of the work," adding that these "works are not only inseparable from their sites – they are not really definable at all apart from them."[1] This is clearly not the case with, for example, most sculpture; with such works, although the site can be aesthetically significant, it is not a part of the work itself.

Environmental art was initiated in the mid 1960s and has continued into present. Throughout this time it has been frequently questioned by individuals concerned about its environmental and ecological consequences. For example, to clear the ground for the construction of *Running Fence*, an 18 foot high, white nylon "fence" running 24 miles across northern California, Christo had to file a 450 page environmental impact report and was required to work closely with local environmental authorities.[2] A more recent and ambitious work by the same artist, *Surrounded Islands* (1983), caused greater controversy. Since the work involved surrounding eleven islands with 5.5 million square feet of pink plastic, environmentalists were concerned about its consequences for the ecology of the islands and attempted to prevent its construction by legal action.[3] Works such as these have even prompted one author to question the morality of environmental art in general. After discussing a wide range of such works, all of which he calls earthworks, Peter Humphrey concludes: "Are earthworks ethical? It is doubtful."[4]

In this chapter, however, I do not consider the ecological, legal, and moral issues raised by these works. Rather I concentrate on only one aesthetic issue about art and nature that arises from the way in which they are related to one another in environmental works. In part because of the way nature becomes incorporated into, becomes a part of, art in such works, it has been suggested by a number of observers that these works constitute something like aesthetic indignities to nature. One artist, for example, views many earthworks "as simple one-sided aesthetic impositions upon nature."[5] Similarly, in a recent discussion of art and nature, Donald Crawford notes that some of such works "forcibly assert their artifactuality over against nature."[6] He adds that although critics usually challenge the environmental impact of these works, one cannot but think that they believe the artists are "engaged in an *aesthetic* affront to nature that goes deeper than the scientific assessment of environmental implications."[7] Following Crawford's way of putting this point, I frame the issue under discussion here by asking if environmental art constitutes an aesthetic affront to nature.

Before discussing this question, it is useful to clarify the issue by noting two points. First, the affront in question is an *aesthetic* affront. I take this to indicate that the affront is generated by the aesthetic qualities of an object, rather than by, for example, its social, moral, ecological, or other such qualities. Crawford, for example, distinguishes a work's "*aesthetic* affront to nature" from its "environmental implications." Thus, although an artwork such as Humphrey's hypothetical *Asian Floodwork*, the object of which "is to show Third-World agriculture under water," would have unacceptable moral and ecological qualities, it would not constitute an aesthetic affront to nature (or to the Third World) on these grounds.[8] Whether or not it would constitute such an affront would depend on its aesthetic qualities, whatever they might be. Second, it is important to note that the aesthetic affront is an affront *to nature*, and not necessarily to normal or appropriate appreciators of the work. The affront in question is by works that "assert their artifactuality over against nature" and are "impositions upon nature." Of course, works that constitute such aesthetic affronts to nature may also affront normal, appropriate, or environmentally concerned appreciators, but such possible affronts are distinct from the issue involved here. Nonetheless, it is true that the *recognition* of an aesthetic affront to nature requires an appreciator, for nature itself cannot recognize the affront. This seems somewhat peculiar, but I take it to be no more so than the fact that I can insult, affront, or impose upon Jones even though Jones may not, and may not even be able to, recognize this.

Environmental art as an aesthetic affront

Given this clarification of the issue, I turn to the question of why anyone would hold that environmental art constitutes an aesthetic affront to nature. The most obvious and intuitive answer is that some of these works are similar in appearance to things that almost everyone agrees are such affronts.[9] For example, Walter De Maria's *Las Vegas Piece* (1969) consists of four, 8 foot by ½ to 1 mile earthmarks in the desert 95 miles northeast of Las Vegas. The work resembles a bulldozer scar on a virgin desert landscape, which is not unexpected, for it was constructed with a bulldozer. Heizer's *Double Negative* is a 50 foot by 30 foot by 1,500 foot double cut in Virgin River Mesa, Nevada, which displaces 240,000 tons of rhyolite and sandstone. It is reminiscent of the results of mining operations, in particular the highwall cuts and skyline notches produced by Appalachian coal mining. Since the aesthetic affronts offered by the latter are significant enough to partly justify reclamation legislation, it is not surprising that some see Heizer's piece as involving a similar affront. Indeed, one critic characterizes such artists as cutting and gouging "the land like Army engineers."[10] Another striking example is Smithson's *Asphalt Rundown* (1969), constructed by dumping a truckload of asphalt down the side of a quarry. It resembles the aesthetic consequences of certain kinds of industrial pollution.

The similarity of appearance of such works to the eyesores produced by industry, mining, and construction is not accidental. Smithson once remarked that the "processes of heavy construction have a devastating kind of primordial grandeur" and that the "actual *disruption* of the earth's crust is at times very compelling."[11] Heizer characterizes himself by saying: "You might say I'm in the construction business."[12] In fact in the early 1970s Smithson contacted industry in an attempt to actualize a proposal for a set of works called *Projects for Tailings*. His vision was to construct earthworks of the millions of tons of waste "tailings" and spoil produced by modern mining operations.[13] It becomes clear that environmental works of art can go further than simply being similar in appearance to the aesthetic affronts of our technological society. They can be virtually identical to them in appearance. Consequently, the claim that such works constitute aesthetic affronts to nature acquires some initial justification. In a way somewhat the converse of the cartoon mentioned in Chapter 9, one can imagine a "leader of industry" responding to a work such as Smithson's *Asphalt Rundown* with the comment: "If my company had dumped that eyesore out there, the government would make me clean it up!"

I suggest, however, that this obvious and intuitive line of justification does not establish its conclusion. This is because I suspect that even if a given environmental work is absolutely identical in appearance to an undisputed

Illustration 3 *Asphalt Rundown*, by Robert Smithson (1969), Rome (*Courtesy of John Weber Gallery*).

aesthetic affront, the former need not have the same aesthetic qualities as the latter. My view is that the aesthetic qualities an object has are only those it appears to have when it is appropriately appreciated and moreover that such appreciation must involve appreciation of that object as the kind of thing it is. Consequently, if two different objects are different kinds of things, they can have very different aesthetic qualities even if they are identical in appearance. This general position is defended in Chapter 5 of this volume as well as in the literature.[14] If it is correct, it means that since it is a work of art, Heizer's *Double Negative*, for example, may be, as one critic describes it, "a deep, majestic double cut," while the skyline notches produced by coal mining, even if identical in appearance, have no majesty whatsoever.[15] Thus, since being an aesthetic affront to nature is a function of an object's aesthetic qualities, the fact that environmental works of art resemble objects that are undisputed aesthetic affronts does not by itself establish that they are such affronts. The aforementioned "leader of industry" is simply not appropriately appreciating these environmental works – he or she is not appreciating them as works of art.

This general position, however, provides a different means by which to support the claim that environmental art constitutes an aesthetic affront to nature. If an object's aesthetic qualities are those given in appropriate appreciation, and if appropriate appreciation involves appreciation of the kind of object it is, then there is a simple way to alter the aesthetic qualities of any object. This is by changing the kind of object it is. Moreover, such changes of the kind of an object and the resultant alterations of its aesthetic qualities can constitute an aesthetic affront to that object – indeed perhaps they necessarily constitute such an affront. The point can be illustrated by examples from the history of art. Consider Leonardo's *Mona Lisa* (1504) and Duchamp's famous mustached and goateed version entitled *L.H.O.O.Q.* (1919). With a few pencil marks and a punning inscription Duchamp changed the work's kind from a Renaissance portrait to a twentieth century Dada statement and in so doing dramatically altered its aesthetic qualities. After Duchamp, the work epitomizes, as one observer – an observer no less than Dali – puts it, the "antiheroic, anti-glorification and anti-sublime aspects of Dada."[16] This is a relatively clear case of an aesthetic affront to the work in question; in fact some critics have used stronger terms, such as a "denigration" of the work.[17] Moreover, I think we find the affront to the work acceptable only because Duchamp used a reproduction of the *Mona Lisa* to execute it. Had he used the original, the impact of the affront would have been greater. Similar affronts to works of art are not difficult to imagine. Consider changing the kind of work the *Guernica* (1937) is, perhaps making it somewhat impressionist, by applying gay, pastel colors to its lighter areas, or, following Monty Python's Flying Circus, changing Michelangelo's *David* (1501–04) into a kinetic sculpture by giving it a moveable right arm. As a kinetic sculpture, the *David* would have radically different aesthetic qualities, even when its moveable arm is at rest in the position originally ordained by Michelangelo. This last point underscores my earlier claim that the affront is not simply a function of the appearance of an object; it is a function of changing an object's kind and thereby altering its aesthetic qualities.

It is in this sense that I think environmental works of art can constitute aesthetic affronts to nature. As noted earlier, a distinctive feature of environmental art is that a part of nature itself is a part of the aesthetic object – the environmental site is an aspect of the work. The environmental artist, sometimes with only a few marks on the land, comparable to Duchamp's pencil marks, changes the kind of thing that this part of nature is. The environmental site is thus changed from being a part of nature to being a part of an artwork and with this change the aesthetic qualities of nature are altered. Heizer, for example, says: "The work is not put in a place, it is that place."[18] And Smithson spoke of his sites as being "redefined in terms of art."[19] If such "redefinition" of kind necessarily involves an aesthetic affront, then environmental art necessarily constitutes an aesthetic affront to nature. However, even if there is no necessity involved, it is not difficult to see many environmental works as comparable to Duchamp's mustached *Mona Lisa* or Monty Python's kinetic *David*, although such works lack the lightness and humor of Duchamp's and the Python's creations. More typically the tone is seemingly set by Picasso's famous remark: "Nature exists to be raped!"[20] In any case, at least some environmental artists seem to have taken his words to heart. For example, Heizer comments: "I find an 18 foot square granite boulder. That's mass. It's already a piece of sculpture. But as an artist it's not enough for me to say that, so I mess with it. I defile... it."[21]

Perhaps "defile" is too strong a word to characterize most environmental art. Nonetheless, the general way in which environmental artists alter nature's aesthetic qualities by turning nature into art does seem to support its being an affront to nature. This is illustrated by Heizer's works such as *Displaced-Replaced Mass* (1969) in which a 52 ton granite boulder is "messed with" by placing it in an excavated depression. It is also evident in works such as Christo's *Surrounded Islands* described earlier and *Valley Curtain* (1971–72), 200,000 square feet of bright orange nylon polyamide spanning a Colorado valley, or Oppenheim's *Branded Hillside* (1969), a "branding" of the land executed by killing the vegetation with hot tar. In such cases nature is "redefined in terms of art" at a cost to its aesthetic qualities such that to speak of an affront, if not a "denigration," is quite appropriate.

Some replies to the affront charge

Rather than further develop the claim that environmental art is for the reasons indicated an aesthetic affront to nature, I now turn to four replies to this charge.[22] One common reply is that environmental works do not constitute affronts because they are more or less temporary. It is true that many works are temporary. Most of Oppenheim's pieces and some of Heizer's earlier ones were constructed such that nature would gradually reclaim the site, and all of Christo's major works are designed to be dismantled after a specific period of time. For example, although *Valley Curtain* took over two years to execute, it was in place for only a few days. However, this is not true of all environmental art. Many pieces by Smithson, Heizer, and others are seemingly intended to be permanent marks on the land, and others are surprisingly durable. For example, although De Maria's *Las Vegas Piece* is only a few inches deep, in 1976 one critic enthusiastically reported: "Six years old and only slightly eroded, the cut appears freshly 'drawn.' "[23]

Nonetheless, even if all environmental works had the relatively short life spans of Christo's pieces, this reply would not be adequate. It might be relevant to the environmental issues that have been raised about works such as Christo's, for the temporary nature of an environmental work can lessen the possibility and the extent of its having a serious environmental impact. However, the issue of an aesthetic affront is of an altogether different nature. An affront is like an insult rather than like an impact, and even if an insult is only temporary, it can yet be as much of and as great an insult as it would be were it permanent. If environmental works of art are aesthetic affronts to nature, they constitute such affronts by the fact of their existence and not by the duration of that existence. Had Duchamp pencilled the mustache and goatee on the *Mona Lisa* and erased it after a few days, he would yet have accomplished his affront to the work. And the consequences of the affront, whatever they are, would have continued to exist after the erasure. Some environmental artists, even though their works are temporary, are well aware of this kind of point. When asked whether the site of *Valley Curtain* remains unaffected by having hosted the work, Christo replied: "Perhaps not. Was Mont-Saint-Victoire ever the same after Cezanne?"[24] If *Valley Curtain* was an aesthetic affront to nature, then the fact of its existence still constitutes the affront.

A second rather common reply to the affront charge is that environmental works are not affronts since by altering nature's aesthetic qualities they improve nature. The original site is seen as having few positive or mainly negative aesthetic qualities and the resultant work of art as an improvement on it. In speaking of Heizer, Smithson, and De Maria, for example, one critic notes that "none of these three artists has opted for sites that are conventionally scenic; their spaces tend to be rather neutral."[25] Smithson in particular is known for his use of "sites that have been disrupted by industry, reckless urbanization, or nature's own devastation," ones he thought could be "cultivated or recycled as art."[26] He remarks: "I'm interested in bringing a landscape with low profile up, rather than bringing one with high profile down."[27] The reply, in short, is that given the barren sites of works such as Smithson's *Spiral Jetty* and *Amarillo Ramp* (1973), Heizer's *Displaced-Replaced Mass* and *Double Negative*, or De Maria's *Las Vegas Piece*, these works do not constitute affronts. They are comparable not to Duchamp's treatment of the *Mona Lisa*, but to improvements made on the work of a third rate hack.

This reply can be met by considering the extent to which the sites in question are in fact natural sites. When they are natural, as the desert sites of Heizer and De Maria, I see no grounds for the claim that they have few positive or mainly negative aesthetic qualities. It is true that such sites are not "conventionally scenic," but, as argued in Chapter 4, "the scenic" is a narrow and parochial construal of nature's aesthetic interest and merit.[28] The desert, for example, has a subtle, quiet beauty of its own, and altering that beauty can be as great an aesthetic affront to nature as altering the aesthetic qualities of conventionally scenic landscapes. In fact I suggest that none of virgin nature is comparable to the work of a third-rate hack – that virgin nature by and large has positive aesthetic qualities. I have developed this suggestion in Chapter 6.[29] If it is correct, the charge that environmental works on natural sites constitute affronts to nature cannot be countered by claiming that the original sites offered little or nothing of aesthetic interest and merit.

On the other hand, some sites such as many of Smithson's are to various degrees less than natural. They are sites such as quarries or spoil dumps which have been, in Smithson's words, "in some way disrupted or pulverized" and "denaturalized."[30] In such cases, to the extent that the site is not a natural site, the environmental work does not constitute an aesthetic affront *to nature*. This is because the aesthetic qualities that are altered by the work are not the aesthetic qualities *of* nature but rather those that have been produced by earlier human incursions. If environmental works on such sites are designed to regain some of nature's original aesthetic qualities, then they may be comparable to an art restorer's treatment of the *Mona Lisa* rather than to Duchamp's. However, more typically, as in Robert Morris' *Johnson Pit #30* (1979), a former strip mine "decorated" with terracing and tarred tree stumps, they are designed to "improve" the site by producing new "artistic" aesthetic qualities. As such, they are comparable to environmental art on any man-made site, such as in the city. Consider, for example, Otto Piene's city sky "ballets." If such works are not aesthetic affronts to nature, it is not because they are not aesthetic affronts, but because they do not have natural sites.

A third, more sophisticated reply to the affront charge draws upon the attempt to identify the works of environmental artists with the works of nature. The idea is that if environmental art and nature are the same kind of thing, then environmental art cannot constitute an aesthetic affront to nature, or at least it cannot do so by changing the kind of thing nature is. The bases for this reply were most clearly articulated by Smithson, who claimed that an "artist's treatment of the land depends on how aware he is of himself as nature."[31] He also urged artists to "become conscious of themselves as natural agents."[32] He characterized his own artistic interest as that of taking "on the persona of a geologic agent where man actually becomes part of that process rather than overcoming it."[33] In this view the changes to nature and its aesthetic qualities produced by environmental works are of the same kind as those brought about by natural processes such as earthquakes or volcanic eruptions. In fact Smithson explicitly compared some of his works – in particular, one with the self-explanatory title *Partially Buried Woodshed* (1970) and the earlier described *Asphalt Rundown* – to the results of natural occurrences such as the Alaska earthquake and the Icelandic volcanic eruptions.[34] The point is that although such natural occurrences change nature and alter its aesthetic qualities, they do not change the kind of thing nature is, and thus, whatever else they yield, they do not result in aesthetic affronts to nature. If environmental works of art are also natural occurrences in this sense, they too can avoid being aesthetic affronts.

Concerning this reply, we may initially note that claiming that an environmental work cannot constitute an affront to nature because it, like nature, is also natural is comparable to claiming that Duchamp's mustached *Mona Lisa* cannot constitute an affront to Leonardo's *Mona Lisa* because it, like Leonardo's work, is also art. In each case, even if we accept the claim of a basic similarity of kind, there seems to be enough difference to make an aesthetic affront possible. Be that as it may, there is perhaps a deeper problem with this reply: it seems, as is said, to throw out the baby with the bath water. If we take the comparison between environmental works and natural occurrences such as earthquakes completely seriously, it becomes difficult to see any point or purpose in environmental art. As we have seen, some environmental artists, including Smithson himself, have elaborated the point and purpose of their art in terms of improvements on nature, of "recycling" "devastated places," or of "bringing up" "low profile" landscapes. But viewing environmental works as natural occurrences undercuts this kind of justification. One cannot consistently hold that these works have the point of improving upon nature and that they yet have the natural purposelessness of earthquakes and volcanic eruptions. Of course, one might concede that such works have neither point nor purpose, that they, like natural occurrences, involve neither improvement nor its lack but only existence and change. It is reported, for example, that Heizer considers the meaning and *raison d'être* of his desert works to be only their "own existence – nothing more and nothing less" – they have "no significance" other than their "presence."[35] This line of thought, however, has its own draw-backs, for it puts environmental art in an uncomfortable limbo between nature and humanity's traditional incursions upon nature. The former, as embodied in natural processes, is characterized by inevitability; the latter, as embodied in activities such as farming and mining, are characterized by purposefulness.[36] However, if environmental artists can claim for themselves neither the inevitability of nature nor the purposefulness of traditional human incursions, then their art may be more similar to simple vandalism than to anything else. If so the charge of being an aesthetic affront is not avoided, for the vandal's marks and alterations, since they are excusable neither by inevitability nor by purposefulness, constitute greater aesthetic affronts than any other. Compare Duchamp's treatment of the *Mona Lisa* with that of a vandal who attacks it with a can of spray paint. Even Smithson once said that he could not accept pointless graffiti on boulders (although it was all right on subway trains).[37]

The last reply I consider attempts to avoid the affront charge by reasserting a purpose for environmental art. However, this purpose is not to improve upon nature by recycling it as art. In fact it is not to alter nature's aesthetic qualities in any way, but rather only to bring out or to make more evident these qualities. In this view an environmental work does not change the kind of thing nature is, turning a natural site into part of an artwork; rather it only "spotlights" the site itself. Thus the environmental piece is construed as a means of displaying and enhancing nature's beauty, and not as an aesthetic affront to nature. Perhaps the genesis of this conception of environmental art is to be found in the traditional idea of art as a mirror of nature or in the idea behind some of Smithson's earlier works which he called "nonsites." These works display in a gallery materials such as rocks and gravel from a natural site and thus call attention to that site without making it a part of the work. Such gallery-bound nonsites do not change nature, but at most only our perception of nature, and thus it is difficult to see how they could be aesthetic affronts to nature. On balance, they are more like providing the *Mona Lisa* with proper lighting than like providing it with a mustache.

Some concluding examples

However, once art moves out of the gallery and directly engages natural sites, the extent to which this reply is relevant is less clear. In fact it seems that whether or not it holds for any particular environmental work depends on the exact nature of the work in question. Thus its strength as a reply must be considered on a case-by-case basis. Consequently, I conclude by examining some examples for which the reply might be thought to hold. Perhaps the best known of works sometimes characterized in this way are some of Christo's such as *Oceanfront Project for Covering the Cove at King's Beach* (1974) and *Wrapped Coast – Little Bay – One Million Square Feet* (1969), each of which involved covering a large area of ocean coastline with fabric. For example, one critic says that the former was "intimately related to the landscape in that the natural forms are accentuated by the fabric – in a sense made more visible, rather than altered or disguised." She concludes: "Concealment by a fabric so malleable, supple, and relatively sheer as polypropylene is thus ultimately revealing."[38] Similarly, another observer notes that "the texture and color of the sand was strangely intensified" by Christo's *Wrapped Coast*.[39] The claim is that these environmental works, rather than alter nature's aesthetic qualities, "accentuate," make "more visible," "reveal," and "intensify" these qualities. However, since the sites are nonetheless covered with fabric, it is not clear that this claim is ultimately convincing.

Somewhat more convincing are similar claims made for the environmental structures of Michael Singer, such as *Lily Pond Ritual Series*, *First Gate Ritual Series* (1976), and *Sangam Ritual Series 4/76* (1976). These works are constructions of materials such as wooden strips, reeds, jute, and bamboo which Singer describes as "clues" to the environment by which nature is made visible.[40] They have also been characterized as "gateways," "reflectors," "accents," and "magnifiers" of the site, and even as "instrument[s] by which each detail of the area might be signaled, might register tellingly on its observer."[41] Such metaphors are somewhat appropriate, for instead of altering or covering their sites, Singer's pieces seem only in a sense to *frame* them, although the frames are internal, skeleton-like frames rather than traditional external frames. Nonetheless, framing of any kind is a recognized artistic method for displaying and enhancing, rather than altering an object's aesthetic qualities. In a similar way it might be said that Christo's *Running Fence internally* frames the landscape through which it passes, although in this case perhaps the frame overwhelms that which is framed.

Another environmental artist comparable to Singer is Alan Sonfist, who art critic Grace Glueck describes as "Nature's boy."[42] Like Singer, Sonfist claims he is "not trying to alter" nature, but "trying to present it"; he wants to create art which makes nature "visible" and "directs" people "to look at nature."[43] To this end Sonfist has developed the idea of "natural phenomena as public monuments."[44] The idea is illustrated by works such as *Rock Monument of Buffalo* (1976–78). In this piece rocks from the local region are positioned such that the work "makes clear in one experience the geology of the entire area."[45] Concerning the issue of presenting rather than altering nature, it is useful to compare Sonfist's piece with works such as Heizer's *Elevated, Surface, Depressed* (1981), a monument in which rocks are mounted on aluminum slabs and positioned according to geometrical rather than geological considerations. By contrast with works such as Heizer's, in Sonfist's, as one critic puts it: "Nature asserts itself as itself."[46]

As this remark suggests, Sonfist takes the concept of environmental art that does not alter, but only displays, nature's aesthetic qualities one step further – to the point of nature itself as art. He claims: "I think nature is art and people have to realize this" and compares himself to Duchamp, saying: "He claimed man-made objects as works of art – I claim natural phenomena."[47] The idea of displaying nature as art is not unique to Sonfist. Consider environmental pieces such as Morris' *Steam Piece* (1967–73), in which vents in a 25 foot site filled the area with steam, or Hans Haack's work with the self-explanatory title, *Spray of Ithaca Falls Freezing and Melting on Rope* (1969). Each of these works does little more than present "natural phenomena" in a natural site; the aesthetic qualities displayed are those of nature itself. The basic idea is given more extensive treatment by Sonfist in works such as *Time Landscape* (1965–78). This work consists of a network of sites throughout New York City, where areas of land have been restored to the way they might have appeared before urbanization. Depending upon the particular site, the land has been replanted with different varieties of trees, shrubs, and grasses in an attempt to recreate pre-colonial landscapes. As one critic says: "*Time Landscape* presents nature in an unadulterated, unmodified state as the fundamental content of the work."[48]

Environmental works such as Sonfist's *Time Landscape* and Morris' and Haack's pieces do little if anything to alter nature's aesthetic qualities, and consequently they avoid constituting aesthetic affronts on these grounds. Such works, therefore, are the only clear cases of environmental pieces that are not aesthetic affronts to nature as that notion is discussed in this chapter. However, they yet present two interrelated problems. First, they suggest the possibility of another kind of aesthetic affront – the affront implicit in the idea that for the aesthetic interest and merit of nature to be recognized it must first be considered a work of art. Second, they pose a general question – the question of why such natural environmental pieces should be considered works of art in the first place. In response to these problems, I suggest, concerning the second, that there are adequate grounds for considering these natural pieces *not* to be works of art. And, concerning the first, if they are not so considered, they do not constitute cases of intimate relationships between art and nature and, therefore, do not raise the possibility of offering significant aesthetic affronts to nature. If this is the case, we are, by these examples of environmental art, re-directed to the issues addressed in the first part of this volume: the important questions of how to recognize and appropriately appreciate nature's own aesthetic interest and merit.[49]

Illustration 4 *Time Landscape*, by Alan Sonfist (1965–78), New York City (*Courtesy of Alan Sonfist*).

Notes

1 Elizabeth C. Baker, "Artworks on the Land," *Art in America*, 1976, vol. 64, pp. 92–6; reprinted in Alan Sonfist (ed.) *Art in the Land: A Critical Anthology of Environmental Art*, New York, Dutton, 1983, p. 75.
2 See Alfred Frankenstein, "Christo's 'Fence': Beauty or Betrayal?" *Art in America*, 1976, vol. 64, pp. 58–61.
3 See "Compromise Proposed in Christo Island-Wrap," *New York Times*, March 20, 1983, p. 54.
4 Peter Humphrey, "The Ethics of Earthworks," *Environmental Ethics*, 1985, vol. 7, p. 21. I am grateful to Humphrey's essay for a number of insightful suggestions.
5 Newton Harrison, paraphrased in Michael Aupling, "Earth Art: A Study in Ecological Politics," in Sonfist, op cit., p. 103.
6 Donald Crawford, "Nature and Art: Some Dialectical Relationships," *Journal of Aesthetics and Art Criticism*, 1983, vol. 42, p. 57.
7 Ibid., p. 56.
8 Humphrey, op. cit., p. 8.
9 Humphrey, ibid., p. 8., for example, motivates his ethical inquiry into earthworks by means of a somewhat similar point:

> ...an earthwork is ethical if and only if what it does to the environment is ethical. This requires us to look at the earthwork as a mark and not only as art. In seeing earthworks in this way, the implication is that earthworks should be ethically judged by the same standard...that all other such marks are judged by – marks like strip mines, dams, and golden arches.

10 Alan Gussow, *A Sense of Place: Artists and the American Land*, San Francisco, Friends of the Earth, 1972, quoted in Robert Smithson, "Frederick Law Olmstead and the Dialectical Landscape," *Artforum*, 1973, vol. 11, pp. 62–8; reprinted in Nancy Holt (ed.) *The Writings of Robert Smithson: Essays with Illustrations*, New York, New York University Press, 1979, p. 122.
11 Robert Smithson, "A Sedimentation of the Mind: Earth Projects," *Artforum*, 1968, vol. 7, pp. 44–50; reprinted in Holt, op. cit. p. 83.
12 Michael Heizer, quoted in John Gruen, "Michael Heizer: 'You might say I'm in the construction business,' " *ARTnews*, 1977, vol, 76, p. 98.
13 For a description of one such proposal and an illustration, see Robert Smithson, "Untitled, 1972" and "Proposal, 1972," in Holt, op. cit., pp. 220–1. For a useful discussion, see Aupling, op. cit., pp. 92–104.
14 Concerning different kinds of artworks, see Kendall Walton, "Categories of Art," *Philosophical Review*, 1970, vol. 79, pp. 334–67; concerning artworks and non-art artifacts, see Mark Sagoff, "The Aesthetic Status of Forgeries," *Journal of Aesthetics and Art Criticism*, 1976, vol. 35, pp. 169–80; concerning artifacts and non-artifacts, see my "Nature, Aesthetic Judgment, and Objectivity," *Journal of Aesthetics and Art Criticism*, 1981, vol. 40, pp. 15–27 (reproduced in this volume, Chapter 5).
15 Baker, op. cit. p. 74.
16 Salvador Dali, "Why They Attack the Mona Lisa," *ARTnews*, 1963, vol. 62, pp. 36, 63–4; reprinted in Barbaralee Diamonstein (ed.) *The Art World: A Seventy-Five-Year Treasury of ARTnews*, New York, ARTnews Books, 1977, p. 326.
17 Ibid., p. 326.
18 Heizer, quoted in Michael Heizer, Dennis Oppenheim and Robert Smithson, "Discussions with Heizer, Oppenheim, Smithson," *Avalanche*, 1970 vol. 1, pp. 48–70; reprinted in Holt, op. cit, p. 171.

19 Robert Smithson, quoted in Lawrence Alloway, "Robert Smithson's Development," *Artforum*, 1972, vol. 11, pp. 52–61; reprinted in Sonfist, op. cit., p. 131.
20 Pablo Picasso, quoted in Andre Malraux, *Picasso's Mask*, trans. June Guicharnaud and Jacques Guicharnaud, New York, Holt, Rinehart and Winston, 1976, pp. 55; the remark is related to environmental art in Mark Rosenthal, "Some Attitudes of Earth Art: From Competition to Adoration," in Sonfist, op. cit., pp. 60–72.
21 Heizer, "Discussions with Heizer, Oppenheim, Smithson," op. cit., p. 178.
22 Some of these replies are suggested by Humphrey, op. cit., pp. 11–18, and Crawford, op. cit., pp. 50–1 and pp. 55–6.
23 Baker, op. cit. p. 80.
24 Christo, quoted in Crawford, op. cit., p. 56.
25 Baker, op cit., p. 75.
26 Smithson, "Frederick Law Olmstead and the Dialectical Landscape," op. cit., p. 124.
27 Robert Smithson, "Conversation in Salt Lake City: Interview with Gianni Pettena," *Domus*, 1972, no. 516, pp. 53–6; reprinted in Holt, op. cit., p. 186.
28 See my "Appreciation and the Natural Environment," *Journal of Aesthetics and Art Criticism*, 1979, vol. 37, pp. 267–75 (reproduced in this volume, Chapter 4).
29 See my "Nature and Positive Aesthetics," *Environmental Ethics*, 1984, vol. 6, pp. 5–34 (reproduced in this volume, Chapter 6).
30 Smithson, quoted in Heizer, Oppenheim and Smithson, "Discussions with Heizer, Oppenheim, Smithson," op. cit., p. 172.
31 Smithson, "Frederick Law Olmstead and the Dialectical Landscape," op. cit., p. 123.
32 Smithson, "Untitled, 1972," op. cit., p. 220.
33 Smithson, "Conversation in Salt Lake City," op. cit., p. 187.
34 Robert Smithson, "Entropy Made Visible: Interview with Alison Sky," *On Site*, 1973, vol. 4, pp. 26–31; reprinted in Holt, op. cit., pp. 192–4
35 Gruen, op. cit. p. 97.
36 See my "On Appreciating Agricultural Landscapes," *Journal of Aesthetics and Art Criticism*, 1985, vol. 43, pp. 301–12 (reproduced in this volume, Chapter 12).
37 Smithson, "Frederick Law Olmstead and the Dialectical Landscape," op. cit., p.127.
38 Sally Yard, *Christo Oceanfront*, Princeton, Princeton University Press, 1975, p. 19.
39 Alan McCullogh, "Letter from Australia," *Art International*, 1970, vol. 14, p. 71; quoted in Carol Hall, "Environmental Artists: Sources and Directions," in Sonfist, op. cit., p. 16.
40 Michael Singer, quoted in Kate Linker, "Michael Singer: A Position In, and On, Nature," *Arts*, 1977 vol. 52, pp. 102–4; reprinted in Sonfist, op, cit., p. 188.
41 Linker, op. cit., pp. 183–4.
42 Grace Glueck, "Art Notes: Auction Where the Action Is," *New York Times*, November 15, 1970, p. D26.
43 Alan Sonfist, quoted in ibid., p. D26.
44 Alan Sonfist, *Natural Phenomena as Public Monuments*, Purchase, New York, Neuberger Museum, 1978.
45 Alan Sonfist, *Rock Monument of Buffalo*, Buffalo, New York, Albright-Knox Art Gallery, 1979; quoted in Jonathan Carpenter, "Alan Sonfist's Public Sculptures," in Sonfist, op. cit., p. 146.
46 Carpenter, op. cit., p. 151.
47 Sonfist, quoted in Glueck, op. cit., p. D26.
48 Rosenthal, op. cit., p. 68. Rosenthal's article provides a useful comparison of Sonfist's *Time Landscape* and Smithson's *Spiral Jetty*.
49 A shorter version of this chapter was presented as part of a symposium on interactions between art and nature at the Xth International Congress for Aesthetics in August, 1984, and appears in the proceedings of that congress. I thank my co-symposiests, Donald Crawford and Dabney Townsend, for helpful comments.

11

THE AESTHETIC APPRECIATION OF JAPANESE GARDENS

The dialectical nature of Japanese gardens

In this chapter I focus on another case of an intimate relationship between nature and art, that which takes place in gardens. As my example, I consider oriental gardens, in particular Japanese gardens. There are, however, different kinds of Japanese gardens with different features, styles, and designs that, accordingly, present different issues concerning aesthetic appreciation. Here I focus only on certain types of Japanese gardens and only on one specific aesthetic issue. The particular types of gardens are often called tea and stroll gardens. Such gardens characteristically possess the kinds of features, styles, and designs that most clearly pose the aesthetic issue in question.[1]

The specific aesthetic issue concerning Japanese gardens can be seen as somewhat of a paradox. I introduce it by reference to my own experience: I find Japanese gardens of the relevant kind very easy to aesthetically appreciate. In them I find myself effortlessly slipping into a state of calm and serene contemplation, marked by feelings of quiet joy and well-being. This in itself is not paradoxical nor even worthy of remark. Japanese gardens are well known, widely acclaimed, and by most accounts explicitly designed for precisely this feature; they typically induce this kind of aesthetic experience. What is remarkable and yields a touch of paradox is that Japanese gardens are in many ways exactly the kind of thing that I, and I think many others, frequently find somewhat difficult to aesthetically appreciate. Let me explain.

In general I, and I assume others, find both pristine nature and pure art relatively easy to aesthetically appreciate. Neither typically causes me appreciative difficulty nor confusion. The relevance of this to Japanese gardens is that such gardens, although also very easy to appreciate, are, of course, neither pristine nature nor pure art. Rather they are classic examples of things "between" art and nature, paradigm cases of the meeting and the mixing of the artificial and the natural. However, such cases of interaction between artifact and nature are frequently found to be somewhat difficult to aesthetically appreciate. This is generally true of what might be called nonartistic cases. For example, many individuals have some aesthetic difficulty appreciating farming, mining, urban sprawl, and other kinds of human intrusions into natural landscapes.[2] Sometimes such intrusions are simply regarded as eyesores, but in many cases, and particularly if the resultant landscapes are not especially unpleasant, the difficulty in appreciation is more a matter of something like aesthetic confusion or aesthetic uneasiness. Moreover, many people have similar aesthetic difficulty appreciating other smaller scale intermixing of the artificial and natural, such as, for example, graffiti on rocks, initials carved in trees, artificially designed plants and animals, and even tattoos. And, again, if the particular instances are not especially unpleasant to the eye, the aesthetic difficulty is more a matter of confusion than condemnation.

More to the point, however, are instances of the meeting and mixing of art and nature that are more clearly artistic, for a Japanese garden is anything but a simple human intrusion into a natural landscape. We might think that artistic cases pose fewer problems for aesthetic appreciation, but this is not necessarily so. In an essay titled "Nature and Art: Some Dialectical Relationships," Donald Crawford distinguishes between traditional harmonious relationships between art and nature and what he calls dialectical relationships.[3] In harmonious relationships art and nature serve as models for one another and in a sense reinforce one another, but do not *interact* with one another. By contrast, in dialectical relationships art and nature are two distinct and frequently conflicting elements whose "interaction is a determining factor in the constitution of the object of appreciation."[4] Crawford suggests that certain kinds of environmental artworks exemplify such dialectical relationships between art and nature, in particular some of the kinds of works discussed in the previous chapter, such as the earthworks of Robert Smithson and the placement pieces of Christo. As noted, many of such environmental artworks are controversial. Crawford notes that they have been attacked on environmental and ethical grounds, and that they pose appreciative problems as well. And, as with the nonartistic cases, the appreciative difficulty seems to involve uneasiness and confusion rather than simply a negative reaction. Thus, environmental artworks that involve Crawford's dialectical relationship between art and nature clearly exemplify what may be called the problem of difficult aesthetic appreciation.[5]

Gardens and especially Japanese gardens are not, of course, earthworks or placement pieces. Do gardens exemplify dialectical relationships between art and nature or rather just the more traditional harmonious relationships? As noted, Crawford characterizes harmonious relationships as ones in which art and nature serve as models for one another. Within the Western gardening tradition there are clear examples of such harmonious relationships. On the one hand, in "French-style" formal gardens, such as some of those at Versailles, harmonious relationships are achieved by art serving as a model for nature. On the other hand, in "English-style" natural gardens, exemplified by what are sometimes called picturesque or landscape gardens, harmonious relationships are achieved by the opposite means, by nature serving as a model for art.[6] However, in the Western tradition there is also another kind of garden, the topiary garden. In such cases, as in environmental artworks, clear dialectical relationships exist between art and nature. In topiary gardens, nature and art are distinct and in a sense conflicting forces and the interaction between the natural and the artificial is constitutive of the object of aesthetic appreciation. And, again as with environmental artworks, topiary gardens are difficult and confusing objects of aesthetic appreciation.

What about Japanese gardens? Do they, like earthworks, placement pieces, and topiary gardens and unlike French formal gardens and English natural gardens, involve dialectical relationships between art and nature? A number of considerations suggest this. For instance, Japanese gardens sometimes include topiary. However, as Josiah Conder notes in his classic study of the subject, "the clipping and carving of trees and bushes into shapes such as mountains, water-falls, boats, and buildings" is "kept within the bounds of moderation."[7] In fact, topiary is increasingly unparadigmatic of Japanese gardens. Nonetheless, dialectical interaction between the natural and the artificial is yet manifest in what Conder calls "a sort of surgical treatment."[8] This treatment, described by another author as the necessity for "pruning, clipping, shearing, pinching or plucking," is an essential part of any Japanese garden.[9] One Japanese authority even notes that if one asks "a Japanese gardener the secret of gardening,...he will hold up his pruning shears."[10] Thus, although they do not serve the representational ends of topiary gardens, the shrubs and trees of Japanese gardens are yet highly artifactualized. And it is not a matter of just a little casual pruning. The vegetation of the Japanese garden is carefully formed and shaped to realize distinctive styles such as in the "ball treatment" and the "fraying treatment."[11] Moreover, the Japanese gardener frequently utilizes complex systems of bending, binding, bracing, and weighting to achieve certain desired effects. Occasionally he or she will even use chemical retardants or thin entire trees by hand – one needle at a time![12]

Moreover, dialectical interaction between the natural and the artificial in Japanese gardens is evident not only in the vegetation, but in the entire landscapes of such gardens. The landscapes themselves are highly artifactualized. For example, Conder notes that in designing the kind of garden known as a hill garden "the hillocks should first be arranged, and then the water channels; the principal rocks and stones are next distributed, and lastly the trees and shrubs are planted."[13] And, as with the vegetation, there are distinctive styles and means of forming, shaping, and positioning hills, rocks and stones, and water. The control of water is especially striking. A popular account of Japanese gardens from the turn of the century notes that Westerners "cannot understand how water, which, like the wind, goes where it wills, has had its bed of white stones built for it, and has been trained, as pet dogs are, to run and tumble and lie down at the will of the master."[14]

The most obvious way in which Japanese gardens manifest dialectical interaction between the natural and the artificial, however, is by, as with many environmental artworks, the careful placement of the artificial within the context of the natural. An essential aspect of the landscape of the Japanese garden is a scattering of artifacts throughout them. These include the paradigmatic lanterns, bridges, and teahouses, as well as pagodas, shrines, wells, arbors, and occasional statuary. One Western introduction to the appreciation of Japanese gardens notes that a superficial contact might unfortunately leave the impression of "a quaint, tinkling medley of little arched bridges, carp ponds, paper lanterns, oddly pruned trees, bamboo blinds, grotesquely jutting rocks, and perhaps a dainty geisha."[15]

The appreciative paradox of Japanese gardens

It is now possible to succinctly state what might be called the appreciative paradox of Japanese gardens. It is this: on the one hand, such gardens lend themselves to easy aesthetic appreciation not unlike each of pure art and pristine nature. But, on the other hand, they are neither pure art nor pristine nature, and moreover they are not even examples of harmonious relationships between art and nature, such as the French formal garden and the English natural garden. Rather, they, like earthworks, placement pieces, and topiary gardens, involve dialectical relationships between the natural and the artificial, and thus they should be similarly difficult and confusing objects of aesthetic appreciation. Consequently, the question posed by Japanese gardens is: why are they not difficult to aesthetically appreciate? Or, to put it another way: how do they manage to so successfully and completely solve the appreciative problem of difficult aesthetic appreciation?[16]

Addressing these questions requires pursuing beyond the intuitive level the theoretical question of why objects that involve dialectical relationships between art and nature pose the problem of difficult appreciation. First, it should be clear that this is not simply because dialectical relationships involve conflicting forces. When conflicting forces are present either in pure art or in pristine nature, they typically give the aesthetic object a dynamic and dramatic quality that makes it an easy and natural focus of appreciation. Rather, I suspect the answer lies not in the conflict between the natural and the artificial, but simply in the differences in their respective natures. I suggest that because of their different natures, each of the natural and the artificial lends itself to somewhat different kinds of appreciation. Moreover, given that in dialectical relationships, as Crawford puts it, "both forces retain their identity as separately identifiable components of the completed work," the resultant object of appreciation is difficult to appreciate.[17] Aesthetic appreciation is difficult because both of the natural and the artificial are independently present, each requires different kinds of appreciation, and thus together they force the appreciator to perform various kinds of appreciative gymnastics. For example, the appreciator may attempt to force either the natural or the artificial component into the appreciative mode of the other or may attempt to achieve some blend of the two kinds of appreciation. However, such appreciative gymnastics are difficult and marked by aesthetic uneasiness and confusion.

What then is the difference between the kinds of appreciation to which the natural and the artificial each lends itself? Much of what is said in Part I of this volume, especially in Chapters 6 and 7, is directly relevant to answering this question. Here I suggest a point similar to that noted there by following up some remarks by Francis Coleman. In delineating what he calls the critical point of view, Coleman opines:

> Suppose we are in the country. We might ask someone who is with us to look at the hills and valleys aesthetically, but never critically...Only if man's efforts are mingled with the products of nature do we speak of criticism or judging or judges...the critic supposes that the object could have been otherwise and that it is as it is because someone designed it to be so. The design rests upon the artist's judgment, and his judgment can be good or bad, faulty or sure.[18]

If Coleman is correct about these differences between appreciating nature and appreciating art, then a significant part of the mode of appreciation appropriate for the artificial is some dimension of critical judgment. Critical judgment is appropriate because the object of appreciation, since it is seen as designed, is seen as something that could have been otherwise. On the other hand, the natural is not seen as designed by an artist whose judgment might have been less than perfect, and thus it is not seen as something that, in the relevant sense, could have been otherwise. Consequently, the mode of aesthetic appreciation appropriate for the natural takes the aesthetic object as given and thus to be, as it were, beyond judgment. If this is correct, then one aspect of the appreciative difficulty involved in appreciating objects constituted by dialectical relationships between the artificial and the natural is the question of the proper role, if any, of critical judgment in such appreciation. It is the appreciative problem of exactly how and to what extent to judge.[19]

If the problem of the proper role of critical judgment is, therefore, part of the difficulty in appreciating objects such as earthwork, placement pieces, and topiary gardens, then the question about Japanese gardens can be put as the question of how they manage to so successfully solve or evade this appreciative problem. Of course, objects of appreciation constituted by dialectical relationships between art and nature may solve the problem in many different ways. However, two general lines of approach suggest themselves. They are by following either the lead of art or the lead of nature. Consider cases such as those discussed in Chapter 10: to the extent that they successfully deal with the problem, most well-known environmental artworks, such as those by Smithson and Christo, do so by following the lead of art in the sense of making the artificial component of the work so dramatic that there is little doubt about the appropriateness of critical judgment. On the other hand, many lesser-known environmental works, such as some by Michael Singer and Alan Sonfist, may follow the lead of nature in the sense of making the artificial component so unobtrusive that, as in the case of pristine nature, the tendency to judge does not arise.[20] I suggest that Japanese gardens successfully solve the problem also by following the lead of nature rather than that of art. However, they do so not by means of making the artificial unobtrusive, but rather by making the natural appear in such a way that the tendency to judge is again averted. In short, Japanese gardens solve the problem of the role of critical judgment by rising above judgment in a way similar to the way nature itself does. Let me explain.

First, consider the aesthetic appreciation of nature. The idea suggested by Coleman's remarks is that in such appreciation we are not confronted by the problem of the proper role of critical judgment because nature rises above such judgment in virtue of appearing as something that could not have been otherwise. This, however, is not to say that any particular bit of nature appears as if it could not have turned out different from how it in fact has. Rather, it is to say that nature in general has a certain kind of look, a look of inevitability – indeed, what we often call a "natural" look. I suggest that Japanese gardens solve the problem of judgment because they, in a way similar to nature, are by and large seen as having this kind of look of natural inevitability. This is a common theme in the literature. For example, even after becoming completely acquainted with the design and construction of Japanese gardens, one commentator describes her reaction as involving the following sentiments:

> What luck, what wonderful luck, these people have! They do not need to make their gardens: Nature has done it for them. It is not that they are so artistic in composing, but only so wise in not changing a single stone or tree from the place in which it was found. And so one feels that the garden *had* to be arranged as it was,...that the lake and the trickling stream and the cascade must have been set there by the Divine Landscape Gardener Himself, and that the beautiful old trees had grown to that precision of shape and loveliness by the help of Nature alone.[21]

As she notes, the general impression is that the garden *had* to be as it is, that is, that it could not have been otherwise. It is thus beyond critical judgment. Moreover, who, after all, would attempt to judge the "Divine Landscape Gardener Himself"?

The fact that Japanese gardens look as if they could not have been otherwise does not mean, however, that they *look just like nature*. This is because, first, in many ways they do not in fact look just like nature. And, second, because if they did, they would not exemplify dialectical relationships between art and nature, but rather simply a kind of harmonious relationship such as, for example, the English natural garden. The English natural garden is more or less a copy of nature and often looks if not just like then at least quite a lot like nature. Rather the point is that a certain kind of look – a look of natural inevitability – is achieved in Japanese gardens and that in the realization of this kind of look such gardens solve the problem of judgment. Relevant to understanding this solution is not only seeing that Japanese gardens have this look but also knowing how it is achieved. I suggest that the key to achieving the relevant kind of look is a kind of idealization aimed at isolating and revealing the essential. In short, the solution to the problem of judgment lies in the fact that Japanese gardens achieve a look of inevitability not by the creation of a simple copy of nature, but rather by the creation of an idealization of nature that attempts to uncover what are taken to be its essential qualities. The guidebook for the well-known Japanese garden in Portland, USA, puts this point simply by saying that the garden is "intended to demonstrate the *essence* of nature."[22] Similarly, one Japanese authority characterizes the Japanese garden as "the continuous endeavor to extract the essence of a stone, a tree, a view" that results in "a celebration of the elementals" and "a glimpse of nature bare."[23] Following Millay, we might say that in the Japanese garden, we, like Euclid, "look on Beauty bare," although the beauty exhibits a natural rather than a geometric facet.[24]

If, in light of the idea of focusing on the essence of nature, we reconsider the ways in which Japanese gardens are artifactualized, the artificial side of

Illustration 5 Willow in the Japanese garden, Portland.

the dialectical relationship between art and nature can be seen in a new light. The artifactualization, although a separate identifiable aspect of the garden, is yet completely subservient to the aim of revealing the essence of nature. Therefore, the artifactualization in Japanese gardens, instead of prompting critical judgment as it seemingly does in many environmental artworks, rather contributes to seeing the garden as something that could not have been otherwise and is thus beyond critical judgment. Consider again the pruning and shaping of vegetation. This should be seen as aimed at what Conder calls "natural prototypes," that is, ideal forms of, for example, the pine, the maple, and the willow, which forms display the essential qualities of the species.[25] As one authority says: the "Japanese gardener can emphasize these qualities, eliminate distracting elements, simplify [a plant's] lines, and thus reveal its true nature to the world."[26] Another commentator puts the point this way: "There is pruning and placing but this results in the revealing of a line which nature itself created and then obscured in its own plenitude."[27] He adds:

> ...this pruning...allows a more natural and, at the same time, more ideal beauty to emerge...[that] is there from the first. It is not created, it is merely allowed to express itself in a louder voice and in plainer terms.[28]

In a similar fashion, the presence of artifacts within the landscape also aims at the accentuation of nature. First, the artifacts are themselves designed and placed, as one source says, "so as to look as accidental and natural as the landscape itself."[29] Second, they by subtle contrast, as another authority stresses, "create within the viewer that feeling of nature heightened which is the salient quality of the Japanese garden."[30]

Conclusion

Thus, in conclusion, I propose the following account of how the Japanese garden, although indeed exemplifying a dialectical relationship between the artificial and the natural, yet deals so successfully with the problem of difficult and confusing aesthetic appreciation that frequently accompanies such relationships: it does so by following the lead of nature in the sense of making the artificial subservient to the natural. It employs the artificial in the creation of an idealized version of nature that emphasizes the essential. It thereby achieves an appearance of inevitability – the look of something that could not have been otherwise – and in achieving this look, it, as pristine nature itself, rises above critical judgment. Therefore, to the extent that the problem of difficult aesthetic appreciation is correctly analyzed as the problem of the proper role of critical judgment in such appreciation, the Japanese garden does not result in difficult and confusing aesthetic appreciation because concerning it the question of critical judgment does not arise. And thus, since it is reasonably pleasant in other ways, the Japanese garden lends itself easily to aesthetic appreciation in spite of involving dialectical interaction between art and nature.

In addition, I note that this account also explains the nature of the aesthetic experience typical of the Japanese garden. As described at the outset, when I experience such a garden I find myself slipping without difficulty into a calm, serene contemplative state, marked by feelings of well-being. Such an aesthetic experience is exactly what one would expect to envelop an appreciator who is in a reasonably pleasant environment that seems as if it could not have been other than it is; moreover, it is the mode of aesthetic experience that quite naturally follows the suspension of critical judgment. However, in spite of the initial plausibility of this account, it may yet be objected that we simply do not suspend judgment, that as a matter of fact we frequently make critical aesthetic judgments of Japanese gardens. I can only respond to this objection by again appealing to my own experience, noting that my aesthetic appreciation of Japanese gardens does not characteristically involve critical judgments. However, perhaps I am aesthetically perverted by my own theory. Thus, I suggest you find a Japanese garden and see for yourself.[31]

Notes

1 A classic Western introduction to the different kinds of Japanese gardens and their different features, styles, and designs is Josiah Conder, *Landscape Gardening in Japan* [1912], New York, Dover, 1964. The oldest source on these matters appears to be an eleventh-century manuscript, *Memoranda on Garden Making*. It is attributed to a Fujiwara nobleman, Tachibana-no-Toshitsuna, and now published as *Sakuteiki: The Book of Garden*, trans. Shigemaru Shimoyama, Tokyo, Town and City Planners, Inc., 1985.

2 I discuss some of the difficulties involved in appreciating farmland in "On Appreciating Agricultural Landscapes," *Journal of Aesthetics and Art Criticism*, 1985, vol. 43, pp. 301–12 (reproduced in this volume, Chapter 12).

3 Donald Crawford, "Nature and Art: Some Dialectical Relationships," *Journal of Aesthetic and Art Criticism*, 1983, vol. 42, pp. 49–58.

4 Ibid., p. 49.

5 I discuss closely related issues in "Is Environmental Art an Aesthetic Affront to Nature?," *Canadian Journal of Philosophy*, 1986, vol. 16, pp. 635–50 (reproduced in this volume, Chapter 10).

6 For philosophically-informed discussion of these distinctions, as well as some of the different aspects of Japanese gardens, see Mara Miller, *The Garden as an Art*, Albany, SUNY Press, 1993, especially Chapter 1, "Definitions, Examples, and Paradigms." Similarly pertinent to these issues as well as to those concerning environmental art are Stephanie Ross, "Gardens, Earthworks, and Environmental Art," in Salim Kemal and Ivan Gaskell (eds) *Landscape, Natural Beauty, and the Arts*, Cambridge, Cambridge University Press, 1993, pp. 158–82; and Thomas Leddy, "Gardens in an Expanded Field," *British Journal of Aesthetics*, 1988, vol. 28, pp. 327–40.

7 Conder, op. cit., p. 7.

8 Ibid., p. 108.

9 David H. Engel, *Japanese Gardens for Today*, Tokyo, Charles Tuttle, 1959, p. 46.

10 Teiji Ito, *The Japanese Garden: An Approach to Nature*, trans. Donald Richie, New Haven, Yale, 1972, p. 140.

11 Conder, op. cit., p. 109.
12 Mitchell Bring and Josse Wayembergh, *Japanese Gardens: Design and Meaning*, New York, McGraw-Hill, 1981, p. 203.
13 Conder, op. cit., p. 132.
14 Mrs. Basil Taylor [Harriet Osgood], *Japanese Gardens*, New York, Dodd, Mead & Co., 1912, p. 7.
15 Engel, op. cit., p. 5.
16 It could be argued that what I call the appreciative paradox of Japanese gardens would not arise (and thus would not require solution) within the framework of Japanese aesthetic appreciation, for such appreciation presupposes a unity of man and nature, of the artificial and the natural, and not the separation of the two that characterizes Western aesthetic appreciation. If so, the paradox is perhaps an interesting example of a quandary generated at least in part by attempting "cross-cultural" aesthetic appreciation. For an excellent short account of how Japanese aesthetics presupposes a "unity and co-identity between man and nature," see Yuriko Saito, "The Japanese Appreciation of Nature," *British Journal of Aesthetics*, 1985, vol. 25, pp. 239–51. Yet, in other essays Saito herself expresses some doubt about the extent to which Japanese gardens exemplify an harmonious unity of man and nature; see Yuriko Saito, "The Japanese Love of Nature: A Paradox," *Landscape*, 1992, vol. 31, pp. 1–8 and especially "Japanese Gardens: The Art of Improving Nature," *Chanoyu Quarterly: Tea and the Arts of Japan*, 1996, no. 83, pp. 41–61.
17 Crawford, op. cit., p. 57.
18 Francis J. Coleman, "What Is the Aesthetic Point of View?," in Francis J. Coleman (ed.) *Contemporary Studies in Aesthetics*, New York, McGraw-Hill, 1968, p. 7.
19 I discuss these issues in more detail in "Nature and Positive Aesthetics," *Environmental Ethics*, 1984, vol. 6, pp. 5–34 (reproduced in this volume, Chapter 6) and in "Appreciating Art and Appreciating Nature," in Kemal and Gaskell, op. cit., pp. 199–227 (reproduced in this volume, Chapter 7).
20 For a good introduction to environmental art, and especially to somewhat lesser-known artists, such as Sonfist and Singer, see Alan Sonfist (ed.) *Art in the Land: A Critical Anthology of Environmental Art*, New York, Dutton, 1983. Ross, op. cit., provides a useful taxonomy of environmental art, sorting works into seven different categories.
21 Taylor, op. cit., pp. 6–7.
22 [no author given], *The Japanese Garden, Portland, Oregon, USA*, Portland, The Japanese Garden Society [no date given], p. 1. The Portland Japanese garden is widely considered to be one of the most authentic gardens outside of Japan. Moreover, it is described in the McFadden guidebook, *Oriental Gardens in America*, as "surely…one of the most beautiful." See Dorothy Loa McFadden, *Oriental Gardens in America: A Visitor's Guide*, Los Angeles, Douglas-West, 1976, p. 198.
23 Ito, op. cit., pp. 197, 139.
24 The idea of the Japanese garden revealing the essence of nature has interesting roots. A noted authority on Japanese art, Donald Keene, relates it to one of the four themes in terms of which he analyzes Japanese aesthetics. He labels the theme "simplicity" and traces it to the Zen preoccupation with the "use of the most economical means to obtain the desired effect." See Donald Keene, "Japanese Aesthetics," *Philosophy East and West*, 1969, vol. 19, pp. 293–306 (the quote in the last sentence is from p. 301). Saito, "Japanese Gardens", op. cit., p. 59, likewise relates the idea to Zen Buddhism, suggesting that by presenting "an idealization of nature" the Japanese garden gives us "a glimpse of this world as it appears to a Zen-enlightened sensibility." She also finds a source in the notion of *kowan ni shitagau*. Literally meaning "following the request," this is the central design principle of the eleventh-century *Sakuteiki* manuscript, op. cit. Originally it was a principle of stone placement, recommending

that a gardener follow the request of the mind (essence, true nature) of an initial stone in the placement of others. However, it was subsequently employed in the placing, designing, shaping, and pruning of other elements of the garden, in each case "following the request" indicated by the true nature of the object in question.

25 Conder, op. cit., p. 108.

26 Engel, op. cit., p. 46.

27 Ito, op. cit., p. 197.

28 Ibid., p.140

29 Conder, op. cit., p. 11.

30 Ito, op. cit., p. 187.

31 A version of this chapter was presented at a symposium on Japanese gardens at the American Philosophical Association Meeting, Portland, Oregon, 1992. I thank my co-symposiast, Donald Crawford and Yuriko Saito, as well as others who have commented on subsequent versions, especially Arnold Berleant, Arlene Kwasniak, Soili Petajaniemi, and Joni Petruskevich.

12

APPRECIATING AGRICULTURAL LANDSCAPES

Traditional agricultural landscapes

Writing in the 1800s, William James gave the following description of a newly established North Carolina farm:

> The settler had in every case cut down the more manageable trees, and left their charred stumps standing. The larger trees he had girdled and killed, in order that their foliage should not cast a shade. He had then built a log cabin, plastering its chinks with clay, and had set up a tall zigzag rail fence around the scene of his havoc, to keep the pigs and cattle out. Finally he had irregularly planted the intervals between the stumps and trees with Indian corn, which grew among the chips; and there he dwelt with his wife and babes.[1]

James noted that his impression of the farmstead was "one of unmitigated squalor" and characterized the scene as "hideous, a sort of ulcer, without a single element of artificial grace to make up for the loss of Nature's beauty."[2]

James's aesthetic judgments may seem harsh; yet had we stood in his place our own judgments may have been similar. The farm James had encountered was both new to the landscape that contained it and unfamiliar to the eye that viewed it. Each of these factors typically contributes to the nature and the intensity of an aesthetic reaction. Consequently, when landscapes molded by agriculture are newly created and unfamiliar, our initial aesthetic reactions to them are frequently as negative as those recorded by James. Fortunately, however, in the time after James's encounter both the landscapes of agriculture and the eyes and minds that appreciated them developed in ways that made possible a more positive aesthetic experience.[3]

By the mid-twentieth century, for instance, the agricultural landscapes of North America were in general quite different from what they were like in James's time. Of course, the exact nature of these more recent agricultural landscapes, and of the fields, farms, and towns that constituted them, varied greatly depending upon places and products. However, whether the landscapes were those of the corn, grain, and dairy farms of the Midwest, the ranches and orchards of the West, or the tobacco, cotton, and truck farms of the South, Southwest, and East, they had all changed dramatically from James's scene of "unmitigated squalor."

For example, at that time the Midwestern agricultural landscape was one of orderly and well-tended fields, bordered by fencerows and windbreaks and punctuated with a wood lot, a rural church, or a country school. Situated at regular intervals across this landscape were relatively tidy farmsteads, with white two-story frame houses and red gambrel-roofed barns. There may have been masonry silos, slat-sided corn cribs, or gable-roofed granaries, and perhaps a chicken house, a milking parlor, or a hog house. If the scene was somewhat cluttered, it was richly diverse – a scattering of chickens, a hay- or bail-stack, various livestock, brightly colored machinery, possibly a windmill, an apple tree, a white picket fence, a dog, a couple of children.

In addition to fields and farmsteads, such agricultural landscapes also contained small rural communities. The point where the roadway intersected the railway had become a main street, dominated by a row of gray and white grain elevators and lined with the shops that served the surrounding countryside. There would have been a grocery, a dry goods store, a hardware store, an implement dealer, a lumberyard, perhaps a creamery, a feed mill, or a remnant blacksmith shop.

These more recent agricultural landscapes were complemented by developments in the sensitivity of many who viewed them. In our aesthetic appreciation of such landscapes we could not but be struck by the orderly pattern, the neat geometry of fields, fencerows, and roadways. However, the most significant focus of appreciation was the farmstead itself. We had learned to perceive it as a pastoral scene. Typically it, just as the fields, was orderly and neat, but even if slightly squalid, it yet was aesthetically rich in appearance and laden with expressive and associative quality. The sights, smells, and sounds of the farmyard, the colors and contrasts of buildings, machinery, and plants and animals, the subtle and intriguing details that permeated the entire scene, all contributed to an appearance of aesthetic diversity and interest. Moreover, it represented the "storybook" farm of many of our childhoods, experienced (often vicariously) as the home of happy hospitable "folks" and "home cooking" and reconstituted in our imagination as the source of both value in our way of life and wholesome food on our tables.

The perceptual and expressive mix that enhanced our aesthetic appreciation of the fields and farmsteads also characterized the rural community. The main street, the storefronts, the row of grain elevators were the charming and picturesque centre of the agricultural landscape, pleasing to the eye in appearance and to the mind's eye in expression. Indeed, at mid-century, one astute observer of the American landscape discussed such communities under the description "The Almost Perfect Town."[4]

In short, by the middle of the twentieth century the fields and farms and farm communities of North America had much of aesthetic interest and merit for anyone who had the imagination and the inclination to appreciate them. However, if this time represented an aesthetic "golden age" for the appreciation of agricultural landscapes, it, like all golden ages, was destined to pass. Agriculture is presently undergoing a revolution and the resulting landscapes are again, like James's North Carolina farm, both newly created and unfamiliar. In confronting the fields, farms, and towns that are being produced by these current agricultural developments, we may find that our aesthetic reactions are more similar to those experienced by James than to those of our own recent past.

The new agricultural landscapes

The last quarter-century has seen technological, economic, and social changes that are revolutionizing agriculture and in doing so bringing about dramatic alterations in the look of the land. The changes are taking place at every level: in breeding and tending, in growing and harvesting, in storing and transporting, in processing and handling, in marketing and merchandising. They involve new machinery, equipment, and buildings on the farm, new crops, chemicals, and fertilizers in the field, and new packages, products, and tastes on the table. In general they are accompanied, on the one hand, by the demise of general farming, the decline of the family farm, and the depopulation of the rural countryside and community; and, on the other, by the growth of specialized "monoculture," the development of mechanized corporate farming, and the birth of "agribusiness." As a result of such changes new agricultural landscapes are emerging that not only are radically different from, but also are rapidly replacing, those we had learned to appreciate.

The exact nature of these new agricultural landscapes is not yet completely clear, but the outline of the new look is not difficult to perceive. In general it involves larger scale and greater uniformity. The primary element of agriculture, the field, is the clearest indicator of these developments. In North America there are now fewer and larger farms than ever before, and the number and size of farms directly effects the size and nature of fields. As agricultural geographer J.F. Hart notes, "few farms, no matter how large they are, have more than eight or ten fields" and "the great majority….are probably subdivided into four or five."[5] Larger farms have resulted in the larger and more uniform fields that are basic to the appearance of the new agricultural landscapes.

Such fields are not simply a function of farm numbers and size, but of technology, especially as manifested in modern farm equipment. The diverse and large-scale machinery of present-day farming makes larger and more uniform fields both possible and necessary. On the one hand, the utilization by agriculture of graders, bulldozers, land levellers, and other earth-moving equipment has made possible "land remodelling" by which the farmer can transform smaller fields broken up by vegetative and topographical features into vast flat tracts of land. On the other hand, such tracts are necessary for

Illustration 6 A traditional farmstead, Minnesota.

Illustration 7 The new agricultural landscape, Saskatchewan.

mechanized farming. Modern planting, cultivating, and harvesting equipment, because of both its size and the subtlety of the operations it performs, requires large uniform surfaces. Such surfaces necessitate less frequent turning, maneuvering, and adjusting of complex machinery and produce regular, even crops that can be handled mechanically at every stage of production. Without large uniform fields, there would be limited use for high-speed tractors, 50-foot cultivators, self-propelled irrigation and spraying systems, and mechanized harvesters and pickers for every crop.

Modern farms and fields complement not only mechanized farming, but also monoculture, the production of a single crop to the exclusion of other uses of the land. Larger farms and more uniform fields are more easily devoted to one crop, whether wheat or corn, beef or turkeys, and doing so is increasingly looked upon as an economic necessity. For the farmer it represents a way to streamline his or her operation and to localize his or her investment of time, energy, and capital. It means capital investment in a smaller range of expensive equipment and personal investment in a smaller range of increasingly complex and technical activities. Moreover, monoculture, when practiced on uniform fields or under the similarly standardized conditions of large farms and in conjunction with the controlled utilization of irrigation, fertilizers, food supplements, and pesticides, produces a uniform crop. Such crops lend themselves not only to mechanized farm equipment, but also to modern methods of produce handling, processing, merchandising, and even consuming.[6]

The movement to a larger scale and a greater uniformity of farms and fields has produced in the new agricultural landscapes what landscape critic J.B. Jackson has described as a "coarseness of detail," a coarse-grainedness characterized by a lack of small and subtle aspects.[7] This is initially most evident in the field itself. Not only have fields become vast flat tracts of land exclusively devoted to a single crop, they have become devoid of many traditional features of rural landscapes. In the quest for large uniform farming surfaces, topographical irregularities such as gullies, washes, sloughs, rises, slopes, and knolls have succumbed to land remodeling. At the same time features once essential to rural life such as wood lots, windbreaks, ponds, fences, country schools, rural churches, and outlying farm buildings are systematically being removed or destroyed. In general such features are themselves deemed unnecessary and occupy land too valuable to remain idle.

The disappearance of traditional features is well illustrated by the case of fences. At one time fencing was so essential to farming that prior to the invention of barbed wire it was thought that the treeless prairies of North America must remain unoccupied for lack of fencing materials.[8] Today however fences and fencerows are all but gone from the new agricultural landscapes. Of the Midwest, for example, the Minnesota Historical Society notes:

> Fences are rapidly disappearing from the American farm…Fence uses change with farm functions. Animals confined to feedlots don't need

> field fences, and fence removal allows larger fields suitable to modern, large-scale machinery. So the once-common barbed wire along highways is now rare.[9]

The changing nature of the agricultural landscapes is apparent not only in a countryside devoid of detail, but also in the farmstead itself. In fact frequently it is apparent in the absence of the farmstead. Geographer Peirce Lewis writes:

> It's no news that the traditional American farm is disappearing. Indeed, in some parts of the country – much of the West, for example – it is already gone...Drive across the country and look around. Farmhouses are empty and falling down, and in the rich cornland of the Midwest a good many have been bulldozed so they wouldn't get in the way of the huge expensive machines of corporate agriculture.[10]

Moreover, even when the farmstead has not completely disappeared, it is frequently totally or partially abandoned. "One harvest of America's increasing productivity," claims the Minnesota Historical Society, "is the abandoned farmstead."[11]

While the totally abandoned farmstead is typically destined for the bulldozer, those only partially abandoned undergo other changes. In many cases only the farmhouse is maintained; the traditional outbuildings – the barn, the corn crib, the chicken house – and the rich detail of the traditional farm – the chickens, the mix of livestock, the windmill – have all disappeared. If the farmstead has become the home of a non-farming family, their passing has probably been gradual; however, if it has become the headquarters for a modern farm, it has probably been more rapid and dramatic. For example, Hart says of the traditional farmyard barn:

> On a modern farm the barn is a relict feature; the hayloft is obsolescent, the threshing floor is obsolete, and who needs stalls for horses? Of course some use can be found for the old building, if it is still structurally sound, but as soon as it starts to deteriorate the best thing to do is to pour kerosene on it and light a match.[12]

Remarking on the modern farm headquarters, another observer notes: "Swept clean of all the usual farmyard clutter of broken and obsolete equipment, chickens, haystacks and manure, the area looks much like an industrial plant."[13]

On today's farmstead the most common feature other than the farmhouse (if there is a farmhouse) is most likely the long, low-profile steel or aluminum shed. These prefabricated, warehouse-like buildings are used in a variety of ways depending upon the nature of the farm: to house and tend stock or poultry, to store various crops, to shelter and maintain machinery. In the new agricultural landscape such featureless metal sheds are to the farmstead what the uniform fields and crops are to the surrounding countryside.

second basic premise, namely, that an aesthetically pleasing environment is to be preferred to one that is not. I think we can avoid being led astray at this point if we consider this premise by means of a somewhat digressional investigation into the fact that it is frequently what is thought of as the *natural* environment that is at issue.

It is quite clear that the naturalness of the environment plays some role in the problems under discussion here. This is suggested by how unsatisfactory we, and certainly environmentalists, must find proposals that certain roadways be "beautified" with durable, life-size, plastic "trees" and "shrubs."[7] It is claimed that such "trees" and "shrubs" are as aesthetically pleasing as the real thing, and in one sense, this may be correct, for if they are good replicas, they will look just like the real thing. Yet most people find such roadside decor quite unsatisfactory, and I suspect that this is at least in part because it is not, in some sense, natural. In light of this, perhaps a proponent of the eyesore argument may claim that even if we aesthetically enjoyed roadside clutter, it still must be cleaned up, for it is not natural. In making this claim such a proponent seems to be shifting the focus of the argument from the aesthetically displeasing nature of roadside clutter to its unnaturalness. However, this shift need not be too great, for perhaps the eyesore argument has a suppressed premise to the effect that roadside clutter is less aesthetically pleasing than the natural environment just because it is less natural. Such a premise might help to meet the claim that roadside clutter can become aesthetically pleasing with camp sensibility. Nonetheless, it seems to commit the eyesore argument to the assumption that there is a positive correlation between an object's being natural and its being aesthetically pleasing. And, although, as noted in Chapter 6, such an assumption may be plausible concerning pristine nature, it is very problematic if the objects in question, as in the case of roadside clutter, are not completely natural to begin with. For example, in both art and craft it is clear that artists and craftsmen make some objects more aesthetically pleasing simply by making them less natural. Consider, for instance, the cabinet maker polishing the natural wood of his or her furniture or the sculptor altering the natural shape of his or her stone. Moreover, even concerning certain so-called natural environments, people sometimes seem to find parts of such environments more aesthetically pleasing when these parts are deliberately made less natural; for instance, there is some evidence to the effect that people aesthetically enjoy artificially thinned forests more than ones left in their natural state.[8] All this suggests that the eyesore argument cannot receive much support from an appeal to naturalness.

Nonetheless, as suggested in the foregoing, naturalness is certainly important to these environmental issues. However, if we shift the focus further toward naturalness, we have a new argument rather than the eyesore argument. This would be an argument to the effect that we must clean up our roadways not because litter and the like is unsightly, but simply because it is not natural. I make a few remarks about this argument because I believe it is what some

The new developments in agriculture that are evident in fields and farmsteads are also reflected in the agricultural community. Larger, more mechanized farms mean fewer people, and the resultant depopulation of the countryside, together with improved transportation, is having a devastating effect on what was once "the almost perfect town." Many small rural communities, like many farmsteads, are simply disappearing, becoming the new ghost towns of North America. The functions they once served are now carried out by what are called "regional centers." For example, although Iowa had in 1964 a higher concentration of "real farms" than any other state, its small towns are now dying."[14] United States Department of Agriculture economist Warren Bailey says of them:

> These towns represent the unfulfilled dreams of the people who went there. They are going the same way as the neighborhood grocery. People want to shop where they have a choice. With air-conditioned cars and good roads, they choose to do their shopping in the cities. Iowa really doesn't have room for more than twelve regional centers.[15]

In the cases where certain towns or "regional centers" are surviving, such communities, due to new farming methods and declining rural population, are undergoing significant changes. Much of the "main street," the center of character and charm in many older communities, is being deserted. The blacksmith shop and the local creamery are long gone, and now the grocery, the dry goods store, and the hardware store are following suit. Those services which are still vital, or newly vital, such as the implement dealer, the chain store, and the fertilizer and pesticide depots, have moved out to the edge of town. Here rows of massive farm equipment, an assortment of storage tanks, and again the ubiquitous metal sheds line the highway strip. Near the vacated main street often only the frame grain elevators remain, and, if the area is prosperous, they are frequently overshadowed by giant round storage towers.

The rows of equipment, the storage tanks and towers, the activity on the highway strip all indicate a rich agricultural region. However, the new agricultural landscapes are landscapes of a mechanized prosperity, a prosperity almost without people. Jackson, who coined the phrase "the almost perfect town," emphasizes some of the ramifications of this:

> It is true we are no longer disturbed by the abandoned one room school or the crossroads General Merchandise; but how will we take the abandoned, more or less modern, high school with monster gymnasium? The abandoned drive-in movie with rows of empty stanchions emerging from the weeds, the abandoned shopping center? We will see them, not only in North Dakota but in Texas and Florida and Kansas and elsewhere."[16]

In spite of the fact that the farm land of North Dakota, Texas, Florida, and

Kansas is rich, in fact in part because it is so rich, the traditional rural community, as the traditional farmstead, is rapidly passing from our agricultural landscapes. As one observer notes, like the small farm, "small towns in America – at least of the Norman Rockwell ilk – are…obsolete relics of a different age. There are no more being built today, and, unless things in America change radically, there never will be."[17]

Difficult aesthetic appreciation and novelty

When perceived in light of the rural landscapes of the recent past, the new agricultural landscapes may initially appear to be aesthetic wastelands. As Lewis points out: "We can study and admire the landscapes our rural ancestors created, but the idea of studying, much less *admiring*, the landscape of modern agribusiness somehow is repellent."[18] If it is repellent, this is not surprising. In the new agricultural landscapes the fields and farmsteads have lost whatever pastoral quality they once had; the rural community is no longer charming and picturesque. The whole is indeed characterized by a "coarseness of detail." In fact the large flat uniform fields devoid of gullies and knolls, fences and wood lots, and, due to modern pesticides, even insects and weeds, may, in spite of their great productivity, strike us as dull or even sterile in appearance. And the impression of barrenness is reinforced by decaying farmsteads and vacated main streets. Moreover, the signs of prosperity both on the farm and in the community, such as the low-profile featureless metal sheds, seemingly add little of aesthetic interest. On the whole, in comparison with traditional agricultural landscapes, the appearance seems that of a "blandscape" rather than a landscape, a "flatscape" of dreary and monotonous sameness.[19] Were James to look upon the new agricultural landscapes, although he could not describe them as scenes of "unmitigated squalor," he might think appropriate the claim that they are "without a single element of artificial grace."

Moreover, if the appearance of the new agricultural landscapes is a factor in rendering repellent the idea of admiring them, their associative and expressive qualities seem doubly so. As noted, the new landscapes are related both to the decline of general farming and the family farm and to the growth of monoculture and specialized farming, together with their increasing dependence on mechanization, irrigation, chemical fertilizers, pesticides, supplements, and the other necessities of the agricultural revolution. The environmental and social price that has been and in the future will be paid for this revolution is by no means clear. Indeed the controversy surrounding this issue is difficult and complex. In spite of lack of clarity, however, it is not hard to find claims such as: "Nowhere is the misuse of resources so evident as in the agricultural landscape."[20] And "more than food rolls off the agribusiness assembly line – rural refugees, boarded-up businesses, deserted churches, abandoned towns, broiling urban ghettoes, and dozens of other tragic social and cultural costs also are products of agribusiness."[21] These and other similar charges may be exaggerated, but there can be no doubt that the environmental and social spinoffs of modern agricultural developments and practices are frequently undesirable.

The aesthetic ramifications of such spinoffs are found in the expressive qualities of the new agricultural landscapes. If the family farm and its rural support community constituted a source of social stability and cultural value in our way of life, and if large-scale mechanized farming has led to their demise, then the mechanized equipment and the vast fields, as well as the abandoned farmsteads and the vacated main streets, express the loss of such stability and value. Similarly, if mixed general farming constituted agricultural practice that was relatively sound from an ecological and environmental point of view, then to the extent that such practice has been replaced by a resource-consumptive dependence on mechanization and irrigation and an environmentally hazardous dependence on chemical agents, we suffer another kind of loss. This is the loss of the vitality, stability, and integrity of our environment, and it is expressed not only by the uniformity of fields and rural countryside, but also by the massive machinery, the fuel and chemical storage tanks, and all the other signs of agricultural prosperity that line the highway strips of regional centers.

If we do indeed find the idea of admiring the new agricultural landscapes repellent, this seems a function of both their blandscape appearances and the expressive qualities we take them to have. Moreover, in light of these factors it may seem that our hesitancy to admire these landscapes is justified, that they are, as James judged the newly established North Carolina farm, completely without grace. However, it must be emphasized that these agricultural landscapes, as the farmstead that James found so distasteful, are a new creation. Thus, as James's farm once did, the new fields, farms, and towns lay somewhat awkwardly on the land, with the unfinished and ambiguous appearance frequently characteristic of the newly created. Moreover, we find ourselves in a position similar to that of James: unfamiliar with the new landscapes and thus with neither eyes nor minds either fully able or willing to appreciate that upon which we gaze. Consequently, we are inclined to judge the aesthetic interest and merit of the new landscapes in relation to that which they have replaced. As James judged his farmstead to have nothing "to make up for the loss of Nature's beauty," so too we may judge the new agricultural landscapes to contain little of aesthetic value to compensate for the loss of the beauty and character of the agricultural landscapes of the recent past.

In order to better understand aesthetic judgments such as James's and our own judgments about the new agricultural landscapes, it is useful to turn momentarily to a more familiar area – the aesthetic appreciation of art. Such judgments seemingly have analogues in the history of art appreciation. Consider the aesthetic reaction of both the public and many art critics to the famous New York Armory show of 1913. The show first brought to the United States the works of various "modern" movements such as futurism and cubism. The products of these movements were judged to be "the most unexplainable and inartistic works," "the weird output of the 'Eccentrics,' " and "the disquieting perpetrations of the art criminals."[22] Duchamp's *Nude Descending a Staircase* (1912) was found to be especially offensive and baffling, being described as a "so-called picture" that "looks like a collection of saddle bags."[23] Concerning this case, as with many similar cases in the history of art appreciation, it is now clear that these works were condemned at least in part because of the novelty of these movements and the unfamiliarity of the spectators with their works. The spectators were thus inclined to judge these works in terms of older appreciative models. The cubist works in particular, being early representatives of that movement, were explicitly compared with "a number of beautiful examples by sane and serious men, whose art only shines the more by its close proximity to this vaudevillian collection."[24]

Concerning cases such as the futurist and cubist works at the Armory show, it may be said that the public and the critics failed to appreciate these works on their own terms. However, this was undoubtedly difficult to do. For example, in part because cubism was a relatively new movement, some of the works in question, such as Duchamp's *Nude*, have a somewhat ambiguous, perhaps not-quite-cubist appearance. In many cases a movement in art must mature in order to fully achieve its distinctive style, yield its most finished and characteristic works, and thereby make manifest how to appreciate any of its works on their own terms. Until a movement reaches such a level of development, it is not clear exactly *what* some of its works *are*, and thus it is difficult to appreciate them for what they are rather than as something else. In such cases the appropriate appreciative models for the products of a movement may not be evident, and inappropriate, often obsolete, models supplied by movements that have been superseded will stand in their stead. Moreover, these features of new movements in art are also present in other cases of aesthetic appreciation, such as that of the new agricultural landscapes. Given the recentness of the agricultural revolution, it is not surprising if such landscapes are similar in certain respects to the futurist and cubist works at the Armory show. Like those movements in 1913, the exact nature of the new agricultural landscapes is not yet clear and they have as yet to fully achieve their distinctive and characteristic look. Thus appreciating such landscapes on their own terms rather than in relation to their own past is understandably difficult.

In order for the works of a new movement in art to be appreciated on their own terms, however, not only must the movement mature, but so too must the eyes and minds of those who view its products. Moreover, for appropriate aesthetic appreciation, this latter development is perhaps the more essential. This is because as spectators mature they are both freed from inappropriate appreciative models and freed to appreciate the newly created works for what they are. They thereby achieve a fuller and truer appreciation of such works. On the one hand, in being freed from inappropriate models, and in casting about for appropriate ones, spectators find how to best appreciate the appearance of new works. For example, the sharp, angular lines and the subdued shades of brown of Duchamp's *Nude* seemingly made it appear harsh, coarse, and dull by comparison with the impressionist works also represented in the Armory show, for "the works of such masters as Manet, Monet, Renoir, Degas" were described as "a relief to eyes and minds tortured by" the works of the cubists, who in turn were judged to be "carpenters" who "in a few weeks...will have to seek places in their real trade."[25] Yet in light of the appreciative models driven home by works such as Picasso's *Guernica* (1937), Duchamp's *Nude* can be appropriately appreciated as a stunning study of form and movement. On the other hand but in a similar fashion, in being freed to appreciate newly created works for what they are, spectators discover the representational and expressive qualities of such works. When appreciated as a cubist study, *Nude* looks very little "like a collection of saddle bags" and much more like a nude descending a staircase. And whatever *Nude* expresses, it is clearly not the case that, as one Armory show critic suggested, "the said expression is one of disordered stomachs or deranged minds."[26]

Appreciating the new agricultural landscapes

In the light of these considerations concerning the appreciation of the works of new art movements, it is fruitful to consider more carefully the aesthetic interest and merit of the new agricultural landscapes. In general, in order to appreciate aesthetically a new landscape, we must, as with the products of new movements in art, view it on its own terms. In part this means that we are not likely to learn to admire a new landscape by extensive comparison with the landscapes that it has destroyed and replaced. The new agricultural landscapes are large in scale and vast in scope. Consequently, we look in vain for the snug little farmstead and the intimate tree- or fence-enclosed field and with frustration and dismay upon that which we find instead. But in so looking we are guided by an inappropriate and perhaps obsolete model. When divorced from this model and appreciated simply in terms of their own appearances, both the new farmsteads and the new fields offer much of aesthetic value.

The farmstead, for example, is now not only neater and cleaner in appearance, but also has an orderly sharpness about it. A sequence of evenly spaced, precisely squared, and perfectly plumbed aluminum structures can have an intensely metallic and boldly geometric elegance all their own. And although the charm of the gambrel-roofed barn and the two-story farmhouse is lost, the low horizontal lines of the new structures, like those of the ranch-style houses that often accompany them, can echo the lay of flat open farm land in a manner reminiscent of Frank Lloyd Wright's "prairie style."[27] However, it is in the fields where the aesthetic impact of the new agricultural landscape is most evident. Here intensity of color and boldness of line combine with scale and scope to produce landscapes of breathtaking formal beauty: great checkerboard squares of green and gold, vast rectangles of infinitesimally different shades of gray, or "immense stripes of sepia and ocher stretching mile upon mile to the margins of the sky."[28] When viewed from high land or a low-flying plane such landscapes match the best of abstract geometrical painting in power and drama.[29] And when standing in their midst one is engulfed by their beauty.

The intensity of color and boldness of line which mark farmsteads and fields are also apparent in the equipment that shapes the new landscape. Consider the following:

> It was still early,...Gigantic silver monsters, half a section long, were crawling across the dark red earth, casting great arcs of water that shattered the white Texas sunlight into a cascade of diamonds. Everything was geometry and primal color – circles and planes and cylinders and parabolas and swirling prisms of transparent light.[30]

These remarks are about a modern self-propelled irrigation system at work, but the described effect seemingly surpasses anything a kinetic sculpture-fountain might achieve.[31] Indeed the appearance of much modern agricultural equipment defies comparison with anything other than sculpture of the grandest and most innovative kind: consider the rich primary colors and the clean, clear lines of any piece of farm machinery, the steady, effortless movement of self-propelled harvesters, pickers, and combines, the massive grandeur of towering silos of shining steel. Such sights can surely dazzle the eye and excite the mind, but to do so modern fields and farms must be viewed without being prejudiced by their own past.

The sights that enliven the farmstead and the field are similarly evident in the rural community. If the highway strip has replaced the main street, it has yet replaced it with what can only be described as a stunning display of color and form. In order to appreciate this display, however, we may have to learn by drawing comparisons with Las Vegas rather than with previous agricultural communities.[32] Nonetheless, if this is so, it again illustrates that, once freed from the appreciative models given by its past, the new agricultural landscapes can be of aesthetic interest and merit. Concerning their appearances, the fields, farms, and towns of modern agriculture are much more than the blandscapes they may initially seem to be.

However, we hesitate to admire the new agricultural landscapes not simply because of what we initially perceive as their blandscape appearance, but also because of what expressive qualities we take them to have. The appearance is enlivened by viewing a landscape on its own terms. In part this means appreciating it without recourse to inappropriate models. But to view a landscape on its own terms also means something more significant than this, and more relevant to its expressive qualities. This, again as with works of art, is to appreciate it for what it is. In the case of any agricultural landscape this means appreciating it as a *functional* landscape.

In general, functional landscapes are those created or molded by humans in order to achieve human goals. Such landscapes are typically deliberately designed to perform the functions necessary for fulfilling relatively important goals. Consequently, functional landscapes are to various degrees both designed and necessary landscapes. In the case of any particular functional landscape, the degree to which it is designed depends in large measure upon the kind of functions it performs and how it performs them; the degree to which it is necessary, upon the necessity of the functions for fulfilling the goal and the importance of the goal itself. In the aesthetic appreciation of any functional landscape it is essential to take into account the degree to which it is designed and is necessary, for these factors are relevant to the determination of what qualities such landscapes have and express.

These considerations are especially pertinent to agricultural landscapes, for such landscapes in general and the new agricultural landscapes in particular are both highly designed and extremely necessary. The kind of functions they perform and how they perform them – using the land itself to produce food and fiber – leads to the creation of highly designed landscapes; and the importance of the goal those functions fulfill – the production of food and fiber – together with the difficulty of fulfilling this goal by alternative means, constitute such landscapes as extremely necessary. Thus, in the appreciation of agricultural landscapes, it is important to consider their designed and necessary nature in order to achieve a balanced aesthetic view.

The aesthetic significance of a landscape being designed is that its aesthetic value depends in part upon how and in particular how *well* it is designed. In the case of many familiar designed landscapes, such as formal gardens, urban parks, and city squares, this point is frequently overlooked or taken for granted, for such landscapes are in large measure designed specifically for aesthetic purposes. Consequently, our aesthetic appreciation of their appearances and expressive qualities itself constitutes an appreciation of how and how well such landscapes are designed. In order to appreciate our everyday functional landscapes in a comparable manner, however, we must explicitly consider their functions and how and how well they are designed to perform these functions. This is in part the point of the much-repeated phrase "form follows function." The forms of all functional objects – buildings, airplanes, and appliances as well as landscapes – must be aesthetically appreciated in terms of how and how well such forms fit their functions. However, the cliché is frequently interpreted too narrowly. With anything functionally designed, not only its form, but much of its aesthetic interest and merit, "follows function."[33]

When considered in this light, the aesthetic appreciation of the functional landscapes of modern agriculture is greatly enhanced for, given the functions they perform, they are in general very well designed. In agriculture, years of trial and error together with the pressures for production have resulted in landscapes that can be appreciated as paradigms of good design – crisp, clean, and uncluttered in appearance and expressive of ingenuity, efficiency, and economy. Moreover, not only the vast open fields and the orderly farmsteads, but also both the machines and the buildings that occupy them, are aesthetically richer when appreciated in terms of how and of how well they perform their functions. Consider again the self-propelled pickers and combines, the mechanized stock buildings, and the new silos and grain elevators of present-day agriculture. These are examples of machines designed to collect, sort, and clean in one continuous flow of activity, of structures designed to provide all the necessities of life, growth, and production, and of storage units designed to follow the natural "angle of repose" of that which they contain. Such machines and buildings not only express the virtues of good design, but possess a style, grace, and elegance seldom exceeded anywhere else. Some observers even claim that in terms of architectural design the landscapes of modern agriculture have achieved a level of sophistication almost unknown in the city and its structures. For example: "What seems to be evolving in our new rural landscape is a form and concept of utilitarian architecture which the city as yet knows little about."[34] Moreover, it is claimed that the machines themselves "are more than mere machines: they are architecture, where form follows function more truly than in most buildings, and with no trace of archness or apology."[35]

Even if functional landscapes, such as the new agricultural landscapes, are very well designed, it is yet important to their aesthetic interest and merit that they be relatively necessary landscapes. A landscape designed to perform functions unnecessary for fulfilling our goals or only necessary for fulfilling trivial or whimsical goals, regardless of how well designed, may express at best playfulness, more typically capriciousness, superficiality, or crassness. Here again we can learn from Las Vegas, but now by contrast rather than comparison. When we regard a landscape as unnecessary, we often cannot take it seriously nor appreciate it as right or appropriate, and thus the possibility of our admiring it suffers accordingly. The importance of perceived seriousness in landscape appreciation and its relationship to necessity is emphasized by geographer Yi-Fu Tuan:

> Landscapes command our respect if we perceive them to be serious… Nature is deemed serious because mountains and plains, forests and grasslands are molded by necessity operating through geological time…As to human environments, those which cater to the biological processes of life are deemed serious. Thus modest homes, country roads, and well-cared fields have traditionally appealed to our sentiments.[36]

We have already noted that in general agricultural landscapes are extremely necessary. They are in Tuan's sense serious landscapes; they "cater to the biological processes of life." In fact many of such landscapes are necessary for our survival. This has always been a dimension of their aesthetic appeal, and unless we judge the new agricultural landscapes to be in this sense unnecessary, they too should command our respect. Thus insofar as "modest homes, country roads, and well-cared fields have traditionally appealed to our sentiments," so now should the elaborate equipment and the vast uniform fields of modern agriculture. They too express the seriousness, rightness, and appropriateness of necessity. Of course, we may doubt the necessity of the new agricultural landscape, and concerning some of its details our doubts may be justified. However, its main trends, its massiveness, mechanization, and monoculture, are probably necessary, and perhaps inevitable, in our modern world. About the latter, for example, agronomist J.R. Harlan writes:

> There are too many people in the world for us to go back to the more complex and more stable agroecosystems...Monoculture is a feature of modern agriculture, and we shall have to learn to live with it; indeed, we might die without it.[37]

Conclusion

There is no doubt that when confronted by the new agricultural landscapes we find their aesthetic appreciation troublesome. The very idea of admiring them may strike us as repellent. Initially we may perceive only a blandscape, dreary and monotonous in appearance and expressive of the loss of both social stability and environmental vitality. However, these landscapes are new and unfamiliar and thus difficult to appreciate on their own terms, that is, without undue comparison to that which they have replaced and as the well-designed and necessary functional landscapes they are.

When appreciated without inappropriate comparison and in light of their functional nature, however, the new agricultural landscapes are infused with new aesthetic interest and merit. With such a landscape, not only is its appearance enlivened, revealing a vibrant richness of color and form, but its expressive character is enhanced by a mixture of good design and serious necessity. Such aesthetic dimensions do not completely obliterate the new agricultural landscapes' sometimes monotonous look and unsavory expressive qualities. Yet these new aesthetic dimensions serve to remind us that these landscapes, like most functional landscapes, are complex, many-faceted, and equivocal objects of aesthetic appreciation. And they demonstrate that our attempts to achieve fuller appreciation of such landscapes are always rewarding and perhaps yield a truer assessment of their actual aesthetic value.

William James saw this clearly and certainly more quickly than many of us. After turning from the new and unfamiliar farmstead that he had just judged in the harshest of terms, he had only to talk with a farmer to revise his opinion.

> I instantly felt that I had been losing the whole inward significance of the situation. Because to me the clearings spoke of naught but denudation, I thought that to those whose sturdy arms and obedient axes had made them they could tell no other story. But, when *they* looked on the hideous stumps, what they thought of was personal victory. The

> chips, the girdled trees, and the vile split rails spoke of honest sweat, persistent toil and final reward...In short, the clearing, which to me was a mere ugly picture on the retina, was to them a symbol redolent with moral memories and sang a very paean of duty, struggle, and success.[38]

His conclusion is relevant to more than that one North Carolina farm: "...wherever there is conflict of opinion and difference of vision, we are bound to believe that the truer side is the side that feels the more, and not the side that feels the less."[39]

Notes

1 William James, "On a Certain Blindness in Human Beings" [1899], *Essays on Faith and Morals*, New York, Longmans, Green and Co., 1949, p. 261.

2 Ibid., p. 261.

3 Throughout this article I follow current usage and refer to agricultural *landscapes* rather than agricultural environments. In doing so, however, I use "landscape" in its geographical sense, which indicates our surroundings or our environment rather than simply a viewed block of scenery or a prospect. On this distinction, see my "Appreciation and the Natural Environment," *Journal of Aesthetics and Art Criticism*, 1979, vol. 37, pp. 267–75 (reproduced in this volume, Chapter 4).

4 J.B. Jackson, "The Almost Perfect Town," *Landscape*, 1952, vol. 2, pp. 2–8.

5 John Fraser Hart, *The Look of the Land*, Englewood Cliffs, Prentice Hall, 1975, p. 74. I introduce and illustrate the issues examined in this chapter by reference to the North American agricultural landscapes and North American agricultural authorities, such as Hart. However, this should not be taken as indicating that these issues are exclusively North American concerns. They are concerns wherever the agricultural revolution is occurring. For example, England's Countryside Commission opens its discussion paper on the new agricultural landscapes of Britain with the following: "There is growing public concern about changes which are taking place in the appearance of the lowland farmed countryside due to modern farming practices." See *New Agricultural Landscapes: A Discussion Paper*, Cheltenham, The Countryside Commission, 1974, p. 1.

6 For discussion of this point, see Jack R. Harlan, "Crop Monoculture and the Future of American Agriculture," in S. S. Batie and R. G. Healy (eds) *The Future of American Agriculture as a Strategic Resource*, Washington, DC, The Conservation Foundation, 1980, pp. 225–50.

7 J.B. Jackson, "The New American Countryside: An Engineered Environment," *Landscape*, 1966, vol. 16, p. 18. This article is an exceptionally insightful and, given its publication date, prophetic discussion of the role of the agricultural revolution in shaping the new agricultural landscapes. It, together with a number of other perceptive and relevant pieces originally published in *Landscape*, is republished in E. H. Zube and M. J. Zube, (eds) *Changing Rural Landscapes*, Amherst, University of Massachusetts Press, 1977, pp. 27–38.

8 See May Theilgaard Watts, *Reading the Landscape of America* [1957], New York, Macmillan, 1975, p. 104. Watts's essays constitute the near perfect guidebook for the appreciation of the rural landscape; concerning agricultural landscapes, see especially "Prairie Plowing Match."

9 [no author given], *Minnesota Farmscape: Looking at Change*, St. Paul, Minnesota Historical Society, 1980, p. 16.

10 Peirce Lewis, "Facing Up to Ambiguity," *Landscape*, 1982, vol. 26, p 20.
11 *Minnesota Farmscape*, op. cit., p 15.
12 Hart, op. cit., p. 136.
13 Jackson, "The New American Countryside," op. cit., p. 19.
14 See John Fraser Hart, "A Map of the Agricultural Implosion," *Association of American Geographers Proceedings*, 1970, vol. 2, pp. 68–9.
15 Warren Bailey, quoted in Nick Kotz, "Agribusiness," in Richard Merrill (ed.) *Radical Agriculture*, New York, New York University Press, 1976, p. 51.
16 Jackson, "The New American Countryside," op. cit., p. 17.
17 Peirce Lewis, "Axioms for Reading the Landscape: Some Guides to the American Scene," in D. W. Meinig (ed.) *The Interpretation of Ordinary Landscapes*, New York, Oxford University Press, 1979, p. 23.
18 Lewis, "Facing Up to Ambiguity," op. cit., p. 20.
19 For a discussion of the nature of "flatscapes," see Ted Relph, "The Landscape of the Conserver Society," in Barry Sadler and Allen Carlson (eds) *Environmental Aesthetics: Essays in Interpretation*, Victoria, University of Victoria, 1982, pp. 47–54.
20 Howard F. Gregor, *Geography of Agriculture: Themes in Research*, Englewood Cliffs, Prentice Hall, 1970, p. 139.
21 Jim Hightower, "Hard Tomatoes, Hard Times: The Failure of the Land Grant College Complex," in Merrill (ed.) *Radical Agriculture*, op. cit., p. 106. For further discussion of the environmental and social spin-offs of modern agriculture, in addition to the other essays in *Radical Agriculture*, see, for example, David Allee, "American Agriculture: Its Resource Issues for the Coming Years," in R. Revelle and H. H. Landsberg (eds) *America's Changing Environment*, Boston, Houghton Miffin, 1970, pp. 56–66; Wendell Berry, *The Unsettling of America*, San Francisco, Sierra Club Books, 1977; Frederick H. Buttel, "Agriculture, Environment, and Social Change: Some Emergent Issues," in F. H. Buttel and H. Newby (eds) *The Rural Sociology of the Advanced Societies: Critical Perspectives*, Montclair, Allanheld Osmun, 1980, pp. 453–88; Committee on Agriculture and the Environment, National Research Council, *Productive Agriculture and A Quality Environment*, Washington, DC, National Academy of Sciences, 1974; P. Crosson and S. Brubaker, *Resource and Environmental Effects of US Agriculture*, Washington DC, Resources for the Future, 1982; Walter Goldschmidt, *As You Sow: Three Studies in the Social Consequences of Agribusiness*, Montclair, Allanheld Osmun, 1978; Wes Jackson, *New Roots for Agriculture*, San Francisco, Friends of Earth, 1980. For a brief overview, see William Aiken, "Value Conflicts in Agriculture," *Agriculture and Human Values*, 1984, vol. 1, pp. 24–7.
22 Respectively, James B. Townsend, "A Bomb from the Blue," *ARTnews*, February 22, 1913; anonymous, "The Armory Exhibition," *ARTnews*, March 1, 1913; and L. Merrick, "Chamber of Horrors," *ARTnews*, March 1, 1913; reprinted in Barbaralee Diamonstein (ed.) *The Art World: A Seventy-Five Year Treasury of ARTnews*, New York, ARTnews Books, 1977, pp. 23–5.
23 Townsend, op. cit., p. 23.
24 Merrick, op. cit., p. 25. In addition to the Armory show case, Monroe Beardsley describes a number of other similar cases from the history of art appreciation, in which, as he puts it, the spectators "failed to see what he [the artist] had *done*, because they approached it [the work] with mistaken assumptions and expectations." He reports that Duchamp's *Nude* was also called "a hurricane in a shingle factory." See Monroe C. Beardsley, *Aesthetics: Problems in the Philosophy of Criticism*, New York, Harcourt, Brace & World, 1958, p. 268.
25 Merrick, op. cit., p. 25.
26 Townsend, op. cit., p. 24.
27 In his early writings Wright observes:

> A building should appear to grow easily from its site and be shaped to harmonize with its surroundings if nature is manifest there...The prairie has a beauty of its own and we should recognize and accentuate this natural beauty, its quiet level. Hence, gently sloping roofs, low proportions, quiet sky lines.

See "In the Cause of Architecture, I," *Architectural Record* [1908], reprinted in Frederick Gutheim (ed.) *Frank Lloyd Wright on Architecture: Selected Writings, 1894–1940*, New York, Duell, Sloan and Pearce, 1941, p. 34.

28 Lewis, "Facing Up to Ambiguity," op. cit., p. 21.

29 It is difficult not to draw direct comparisons with some of the works of painters such as Kenneth Noland, Barnett Newman, Josef Albers, and Burgoyne Diller – even though such comparisons may import appreciative models as inappropriate as those given by our more traditional agricultural landscapes.

30 Lewis, "Facing Up to Ambiguity," op. cit., p. 21.

31 However, given Lewis's description, some of the works of Naum Gabo and of certain other constructivism sculptors achieve a somewhat similar effect, although typically on a smaller scale. For example, some aspects of the description are reminiscent of Gabo's *Spiral Theme* (1941) even though the sculpture is not a fountain, not kinetic, made of plastic, and only inches in height. Again, however, such comparisons, as those with abstract painting, must be considered with caution.

32 See Robert Venturi, Denise Scott Brown, and Steven Izenour, *Learning from Las Vegas: The Forgotten Symbolism of Architectural Form*, Cambridge, MIT Press, 1977.

33 Architect Louis H. Sullivan, to whom the cliché is typically attributed, did not give it a narrow interpretation. That "form ever follows function," he describes as "the pervading law of all things...of all true manifestations of the head, of the heart, of the soul," adding that "shape, form, outward expression, design or whatever we may choose...should in the very nature of things follow function." He continues, "when the known law, the respected law, shall be that form ever follows function;...then it may be proclaimed that we are on the high-road to a natural and satisfying art,...an art that will live because it will be of the people, for the people, and by the people." See "The Tall Office Building Artistically Considered," *Kindergarten Chats and Other Writings* [1918], New York, Wittenborn, 1979, pp. 208–13.

34 Jackson, "The New American Countryside," op. cit., p. 19.

35 Lewis, "Facing Up to Ambiguity," op. cit., p. 21. If we were to follow Sullivan, we might put part of the point of this paragraph by saying that well-designed functional landscapes, buildings, and machines, such as those of modern agriculture, may be appreciated as the products of "a natural and satisfying art,...an art...of the people, for the people, and by the people."

36 Yi-Fu Tuan, "Visual Blight: Exercises in Interpretation," *Visual Blight in America*, Washington, DC, Association of American Geographers, 1973, p. 26.

37 Harlan, op. cit., p. 232. Of course, not all authorities agree with the position represented by Harlan. For an overview of some alternatives to large scale monoculture, see Richard Conviser, "Toward Agricultures of Context," *Environmental Ethics*, 1984, vol. 6, pp. 71–85. Moreover, the actual aesthetic ramifications of the line of thought of the last part of this section depends upon what we take our human goals to be and how we evaluate their importance. The accepted goal for the functional landscapes of agriculture may be simply to ensure the survival of as many human beings as possible for as long as possible. However, we may recognize more comprehensive systems together with different goals and evaluate these goals as equally or as more important. For example, one such goal might be, following Aldo Leopold's "Land Ethic," "to preserve the integrity, stability, and beauty of the biotic community" (including, I assume, at least some human beings). See "The Land Ethic" [1953], *A Sand County Almanac with Essays on Conservation from Round River*, New York, Random House,

1974, p. 262. When perceived in light of such considerations, many agricultural landscapes may again reveal rather unsavory expressive qualities similar to those discussed in the third section of this chapter. Although I am sympathetic to these considerations, they cannot be pursued here. Within the context of this chapter my point is simply that given the goals that many of us in fact accept for the functional landscapes of agriculture and given the importance we grant to these goals, we should find such landscapes more expressively appealing than we typically do.

38 James, op. cit., p. 262.

39 Ibid., pp. 260–1.

13

EXISTENCE, LOCATION, AND FUNCTION: THE APPRECIATION OF ARCHITECTURE

Architecture and art

What of general aesthetic interest and significance can be said about the appreciation of architecture, an art marked by works, movements, traditions, and theories possibly more diverse than those of any other major art form? Perhaps preliminary progress is best made only by noting some obvious ways in which architecture as an art differs from many other forms. These differences mean that the aesthetic appreciation of architecture poses certain challenges not typically present in the appreciation of other arts. Such challenges not only shape the nature of appropriate appreciation of works of architecture, but also help to make such appreciation especially rewarding.

Initially it is important to again stress what has been suggested in a number of previous chapters: that aesthetic appreciation involves more than simply either passive contemplation of pleasing form or spontaneous delight in sensuous surface. Essential to aesthetic appreciation is active engagement, involving cognitive and emotional interaction between the appreciator and the object of appreciation. An important aspect of this engagement is a kind of dialogue between appreciator and object in which the latter explicitly or implicitly poses certain questions or problems and the former finds the answers or solutions. Such finding of answers or solutions typically takes the form of coming to realizations about the nature of the object of appreciation. This process of realizing is at the heart of aesthetic appreciation; it employs the imagination so as to produce that unique combination of admiration and awe that is central to aesthetic experience.[1]

The questions that are posed by a work of art vary according to its kind, for different art forms present different problems to be solved. Consider a representational painting, for example, a small sketch by Group of Seven artist Tom Thomson such as *Autumn Foliage* (1916).[2] Such a work constitutes a straightforward case, for with representational works the initial and obvious problems to be solved typically concern what is represented. In *Autumn Foliage*

Illustration 8 Autumn Foliage, by Tom Thomson (1916) (*Courtesy of the National Gallery of Canada, Ottawa*).

we confront a small but vibrant mass of fiery color, but even without reference to the title we easily experience the work as a landscape. We see it as the red and yellow autumn foliage of eastern hardwoods against a background of blackish blue water, distant stands of dark conifers, and an almost turquoise fall sky. Such realizations come almost without effort to any appreciator familiar with early twentieth-century painting. However, even relatively simple representational works pose further problems that offer the imagination somewhat more exercise, typically questions concerning how and why the work is executed as it is and what the appreciative consequences are of the particular execution. For example, if we consider the question of the significance of Thomson's dimensions, imaginative contemplation rewards us with the realization of the extent to which its small size contributes to its power. Its borders confine the fiery mass of color such that it struggles to escape from the surface, and we experience a work that glows and flares like embers enclosed in a pressurized chamber.

To be or not to be: Hamlet and Tolstoy

In contrast to art forms such as representational painting, works of architecture typically pose some larger if not deeper questions. Perhaps the most basic of these is captured by the opening line of Hamlet's most famous soliloquy: "To be, or not to be: that is the question."[3] Hamlet's question is whether it might not be better – "nobler in the mind" – not to exist, and this is a question that seldom arises when we view conventional works of art. It is difficult to imagine a work such as *Autumn Foliage* posing such a question concerning its existence. Of course, this is in part because *Autumn Foliage* is a small masterpiece and as such seems to fully justify its own existence. However, even were this not the case, the question does not easily arise for such works. Perhaps this is for reasons similar to those that apparently move Hamlet toward the conclusion that it is better to be. The alternative, to die, is to be no more, or worse, and "there's the rub," to suffer other ills "that we know not of" – the fact that the alternative is unknown "does make cowards of us all."[4] Similarly, as we contemplate *Autumn Foliage*, even were it not a masterpiece, it would yet be the case that were it not there, we would contemplate only the blank wall of the gallery or worse some other work that we know not of. Apparently, with such art forms, the question of existence arises only if the work is so bad that almost anything else, even a blank wall, would yield a better aesthetic experience.

Concerning architecture, however, Hamlet's question is very much alive. Consider Philip Johnson's American Telephone and Telegraph (AT&T) Building, monumental, granite clad, standing boldly on the street edge, and capped with what has been called a "Chippendale highboy top."[5] As we contemplate it, even if we believe it a masterpiece, we cannot help but consider the question of its existence; it is forced upon us. We ask ourselves: Might it

Ilustration 9 The AT&T Building, by Philip Johnson/John Burgee (1980–83), New York City (*Courtesy of American Telephone and Telegraph*).

not have beeen better for this not to have existed? Might it not have been better for the place, for the skyline, for the city, for the world? Such questions are forcefully posed by the building itself, but not simply because of the work's monumental size or its controversial postmodern style. With architecture such questions are posed in part because, unlike with either *Autumn Foliage* or Hamlet, the alternatives to a work's existence are typically not either nothing or the unknown. Had the Johnson AT&T Building not been constructed, there would not have been nothing, there would not have been a blank space analogous to a blank wall. Rather there would have been another work, either the previous building or a different work of architecture, or if not that, then at least the lot or the city block with its own aesthetic features. Moreover, such alternatives are not unknown; in contemplating them we are not in a position analogous to contemplating what death might hold or what might go on the blank gallery wall. We can know exactly what the previous building was like, know roughly what the city block would be like, and have a good idea of what kind of alternative architectural work might have existed had the one we confront not existed.

Whether or not the question of its own existence is posed by a work of art has significant ramifications concerning its appreciation. Once the question is broached, attempting to answer it brings realizations that are central to our aesthetic experience of the work. However, if a work does not pose the question, then this is not the case. In appreciating *Autumn Foliage* it is appropriate to consider whether or not the work might have been improved by a somewhat less turquoise sky. Such consideration can bring us to the realization that the turquoise is just what is wanted. A stronger blue would have not only been less true to a northern autumn sky, but also contrasted too much with the fiery reds and yellows, and thus would have resulted in a less subtle work. Such realizations enhance our appreciation of the work and are a central part of it. In contrast, considering whether it might not be better for *Autumn Foliage* not to exist and therefore to contemplate the blank wall of the gallery or have its space on the wall filled with, say, *Guernica* (1937) is quite irrelevant to our appreciation of the work, if not just absurd or at least aesthetically perverse. Such consideration is not a proper part of our appreciation of that work; it is rather just one kind of lack of attention to it. As noted, a work such as *Autumn Foliage* would pose the question of its existence only if it were excruciatingly bad, and in that case perhaps lack of attention is the proper response; but this is yet lack of attention to the work, not a part of its appreciation.

We can, of course, consider the AT&T Building's Chippendale top in the same way we consider *Autumn Foliage's* turquoise sky, asking, for example, if perhaps a domed top or a simple flat top might not have been better. Doing so brings realizations that enhance our appreciation of that work and are a proper part of that appreciation. However, unlike *Autumn Foliage*, the AT&T Building forces Hamlet's question upon us and thus opens the door to another level of aesthetically relevant consideration. As we contemplate the work, we ask ourselves what if, for example, Johnson had not created it, but instead of it another modernist skyscraper, such as the classic Seagram Building, which he earlier collaborated on with Ludwig Mies van der Rohe. We imaginatively consider the alternative aesthetic effect that such a clean and crisp, glass and steel structure might have had and the ways in which it might have blended rather than contrasted with a skyline dominated by similar structures. Since the AT&T Building itself poses the question of its existence, such consideration of alternatives to its existence are a proper and central part of its aesthetic appreciation. Moreover, the same is true of our consideration not only of alternatives to its existence, but also of its simple nonexistence. This is easier to see concerning works in natural landscapes. Consider Frank Lloyd Wright's famous houses such as Falling Water and Taliesin West.[6] The former grows out of a rocky Pennsylvania ravine scattered with poplar and birch while the latter sprawls across a stretch of Arizona desert dotted with sagebrush and cactus. In each case the work candidly asks us to consider the fact of its existence, and thus our experience of the work rightly involves imaginative contemplation of the landscape without the work. And this contemplation is a central and proper part of our appreciation of the work; such appreciation is typically deepened and enriched by the realizations it initiates.

The central place of the question of existence in the appreciation of architecture raises concerns about how to address it. As noted, the realizations gained from attempting to answer it are as significant as are those gained from answering questions about features more internal to the work, such as questions about the rightness of *Autumn Foliage*'s turquoise sky or of AT&T's Chippendale top. This in turn suggests that the appreciation of architecture is necessarily a more broadly based and less insular experience than is the appreciation of some other art forms. The borders between architecture and the world in general and the world's aesthetic issues and its ethical, social, political, and even economic issues are not as hard and fast as are those between, say, landscape painting and such issues. Moreover, given the central place of the question of existence in the appreciation of architecture, it seems that there should be some core idea in terms of which to answer it. But finding this idea initially appears exceedingly difficult in light of the lack of hard and fast borders between architecture and the concerns of the world in general. In short, many more kinds of consideration seem relevant to the question of the rightness of the AT&T's existence than to the question of the rightness of *Autumn Foliage*'s turquoise sky. But in addressing the former question not absolutely everything can be given attention and it is difficult to see what should be deemed essential.

Concerning this issue, it is illuminating to consider a classic discussion in which Hamlet's question is asked about art in general. In one of the most remarkable passages in the history of aesthetics, Tolstoy launches an aggressive attack on art, calling on it to justify its own existence.[7] His prime example is a work somewhat comparable to a work of architecture such as the AT&T Building in that it has monumental size and involves extensive utilization of resources; it is an opera that Tolstoy describes as "one of the most gigantic absurdities that could possibly be devised."[8] He directly challenges the existence of such works, asking for what purpose and for whom are they created. Moreover, his concern is not simply with large and imposing works but with art in general; it focuses on the idea that for "the production of every ballet, circus, opera, operetta, exhibition, picture, concert, or printed book, the intense and unwilling labor of thousands of people is needed at what is often harmful and humiliating work."[9] Of course, work in the present-day designing and building disciplines and trades is probably not comparable to the labor of workers in Tolstoy's Russia, but it is yet possible to motivate the question of existence posed by architecture in a way somewhat reminiscent of Tolstoy. And this is not surprising in light of the previously noted openness in the borders between architecture and the world at large.

What is especially relevant to architecture, however, is Tolstoy's approach to answering the question of existence. He recognizes that what is needed is, as noted above, some core idea that can determine, given Tolstoy's interests, "what is good, useful art – art for the sake of which we might condone such sacrifices as are being offered at its shrine."[10] And significantly he finds his answer by casting the question in functional terms, asking "for what and for whom" does art exist and then providing a theory of art that gives art a functional role in human life. As is well known, in Tolstoy's theory of art, real, non-counterfeit art – the only art that should exist – has the function of communicating feelings between people, and "good, useful art" is the art that unites humankind by communicating feelings of love. We need not, of course, accept Tolstoy's whole theory of art to find relevance in his general line of thought concerning the question of existence. The relevance is in the realization that it is necessary to cast our question and its answer in functional terms. And although doing this would require a functional theory such as Tolstoy's were we interested in art in general, since our concern is only with architecture, it does not. This is because, unlike many art forms, architecture is by its nature a functional art. Thus, when Hamlet's question is posed by works of architecture, Tolstoy's implicit answer is to emphasize, rather than ignore, their functionality. I return to this answer, but first it is useful to expand and deepen the question to which it is addressed.

Here I stand: to fit or not to fit

If Hamlet's question is an apt way to indicate the first challenge offered by works of architecture, a second can perhaps be marked by Martin Luther's famous affirmation of his faith. In 1521 at his second hearing at Worms, Luther, accused of heresy and threatened with excommunication and death, is reported to have concluded his defense with these words: "Here I stand. I cannot do otherwise, God help me. Amen."[11] Luther, of course, was defending a theological position, but it is the way in which he did so that is evocative. This affirmation of his position has a firmness and finality that suggests the firmness and finality with which works of architecture, especially large and massive works such as the AT&T, occupy their physical positions. The AT&T Building, as noted, is monumental in size, encased in granite, and stands boldly at the street edge. It is not set back and it is not light; it is imposing and solid. With only a little imagination it is difficult not to perceive it as strongly affirming its physical position, as an exemplification of "Here I stand. I cannot do otherwise." But, of course, a powerful affirmation always offers the challenge of whether or not this is how it should be, or, in this kind of case, *where* it should be. Thus, the second question posed by works of architecture elaborates the first. They pose not only the question of to be or not to be, but also that of to be or not to be *here*?

The "here I stand" question is most obviously raised by works such as the AT&T, works of size and of imposing nature, especially skyscrapers, office towers, and large luxury hotels. For example, modernist towers such as the Seagram Building, the Union Carbide Building, and the Chase Manhattan Bank each pose the question on the skyline of New York City.[12] In a similar way, the affirmation is made by cathedrals and temples, castles and capital buildings. Notre Dame de Paris, the Parthenon, the monastery of Mont St. Michel, and any decent US state capital building each in its own way proclaims "Here I stand." However, it would be a mistake to think that this issue is raised only by substantial structures directly proclaiming their own "hereness." Wright's Falling Water raises the question in virtue of how it is built into its site, embodying a design absolutely dependent on that site. Likewise, at Taliesin West, features that are cited as evidence of Wright's so-called "organic" style – such as the use of natural materials, the rough finish of the timbers, and the way the structure sits on its stepped site – each give the work a special relationship to that particular site, a relationship underscoring the importance of where it is. Other kinds of works pose the question of their location in yet other ways. Le Corbusier's Villa Savoye perches on the landscape, poised and alert, like an alien object, the tenuousness of its relationship to its site not so much making a proclamation as directly asking the question. It seems to say not "Here I stand" but something like: "Why am I here?"[13]

Concerning a work's location, the comparison with other art forms is again illuminating. A painting such as *Autumn Foliage* has, of course, a location. As a physical object, it must be in some place or other. But as a work of art, it does not raise the issue of its location, it does not proclaim, or pose the question of, where it is. Calling its location a *site* would be at best misleading and considering the work in its location, whatever that is, is not a part of its appropriate appreciation. Of course, in attempting to appreciate such a work we may be forced to consider its location, if, for instance, it detracts from the work, say, with noise or poor lighting or conflicting color. But such consideration, and the realizations arising from it, are not proper parts of our appreciation of that work, but rather only distractions from it. The only relevance such consideration might have to the appreciation of the work would be in ways such as initiating moving the work to a "neutral" location and thereby making its full appreciation possible. However, there is one sense in which works such as *Autumn Foliage* have a proper place or "site." This particular work is located on a wall in the National Gallery of Canada, in one of the rooms of early Group of Seven and related paintings, alongside a number of other small Tom Thomson landscapes; and this is a, if not the, proper place for it to be. But, although this placement of *Autumn Foliage* may be proper and may even contribute to its appreciation, this does not make its actual physical location directly relevant to that appreciation. Its physical location indicates its proper conceptual and art-historical placement and it is the realization of the latter, not the former, that is a proper part of its appropriate appreciation.

Not all works of art, however, are location independent in the sense exemplified by paintings such as *Autumn Foliage*, and it is revealing concerning architecture to consider some art forms said to be "site specific." Site specificness is claimed to be a feature of some large sculptural works, of most earthworks, and of what are called placement pieces. Consider Richard Serra's controversial *Tilted Arc* (1983), a 12 by 120 foot curved sheet of oxidized steel cutting across a public plaza.[14] Reaction to the piece resulted in a hearing concerning whether or not to relocate it. In defense of his piece, Serra claimed: "I don't make works that can be relocated or site adjusted. I make works that deal with the environmental components of given places. The scale, size, and location of my site-specific works are determined by the topography of the site." He concluded: "My works become part of and are built into the structure of the site…To remove *Tilted Arc*, therefore, would be to destroy it."[15] Indeed, in the hearing the relationship of the Arc to the plaza was compared to that of a painting to, not "its" wall, but its canvas. As noted in Chapter 10, similar claims are made for earthworks and other placement pieces. For example, in reviewing classic works such as Robert Smithson's *Spiral Jetty* (1970) and Michael Heizer's *Double Negative* (1969–71) and *Complex One* (1972–76), Elizabeth Baker, editor of *Art in America*, points out that the sites of such works "become places as vivid as the works themselves – they become concertized, identifiable, specific locales" and that therefore the "appearance of that place becomes a part of the content of the work" to such an extent that the works "are not only inseparable from their sites – they are not really definable at all apart from them."[16]

If such works are "part of the structure of the site" and the site is "part of the content of the work," there are important consequences concerning aesthetic appreciation. It means that to appreciate the work itself is to appreciate its location, that is, to come to realizations about the relationship, the fit, as it were, between the part of the work the artist has added and the part of the work that is the site of that addition. In short, to appreciate such a work is to appreciate its fit, coming to realizations about whether or not it fits and, if it fits, how it fits. For example, concerning Heizer's monumental desert work, *Complex One*, Baker records the following realizations:

> The frontality of *Complex One* can be seen as projecting a curiously pictorializing effect on a portion of the landscape: it is not just that the presence of the piece re-presents or differentiates this particular place; it is also that as a distant, planar picturelike entity the work tends to crystallize and make a panorama of a certain lateral sweep of land that frames the frame of the piece. [17]

In a somewhat similar manner, essentially involved in the dispute over *Tilted Arc* was the realization that the additions involved in Serra's works "often restructure, both conceptually and perceptually, the organization of the site."[18] In appreciating *Tilted Arc*, viewers came to the realization that part of what the work does is "to alter and dislocate the decorative function of the plaza, to redefine the space, to change the viewer's experience of that plaza."[19] In fact much of the objection to the work was based on this realization and the opinion that this was essentially "to destroy the plaza's original artistic concept."[20] In short, some viewers appreciated the work, including the fit of the arc with its location, and they did not like that fit.

It is not clear that works of architecture relate to their locations in the very strong way in which works such as *Tilted Arc* and *Complex One* do. It seems that the location of a work of architecture is not always, in any case, "part of the content of the work." Nonetheless, works of architecture are certainly closer to these kinds of works than they are to works such as *Autumn Foliage*, and thus the way in which the former kinds of works are appropriately appreciated is illuminating concerning architecture. Thus, in appreciating architectural works we must appreciate the relationship of the structure to its site as a part of the total experience. The fact that works of architecture pose the "Here I stand" issue is in itself sufficient to make the fit of a work to its site a central feature of its proper appreciation. Nowhere is this clearer than with works such as Wright's "organic" style houses. Taliesin West "differentiates," "re-presents," and "crystallizes" its desert landscape site in a way remarkably similar to that which Baker attributes to Heizer's *Complex One*. And the relationship between Falling Water and its ravine is as intimate as that between any earthwork and its site. Architect and historian Kenneth Frampton says of Falling Water: "Its fusion with the landscape is total, for…nature permeates the structure at every turn."[21] Here one is tempted by the claim that the ravine is indeed part of the work, but even if that does not withstand scrutiny, it is yet obvious that we do not fully appreciate, if appreciate at all, the work of art called Falling Water unless we appreciate the rocky treed ravine and the way in which the structure is designed for and built into it. Here the fit of building and site is an essential, if not the essential, dimension of the work.

It is a long way, however, from Falling Water to modernist towers such as the Seagram Building, the Union Carbide Building, and the Chase Manhattan Bank. Although such structures clearly raise the question of their location, it is initially less clear that their proper appreciation involves the appreciation of their fit with their sites. Perhaps they are after all more like *Autumn Foliage* than like Falling Water. However, the flaw in this view is made clear by further comparison with examples such as *Complex One* and *Tilted Arc*. Each in its own way demonstrates a different facet of the proper appreciation of such structures. On the one hand, much as does *Complex One*, and, for that matter, Taliesin West, a tower such as the Seagram Building "differentiates," "represents," and "crystallizes" its site. This was especially evident when this particular work was constructed in 1958: it was Mies van der Rohe's first New York building and, with 38 stories of glistening bronze and brown glass, the largest of its type in the world. And it was set back 90 feet from the line of the street, both as a way of creating a plaza on the site and, it is said, "as a complement to the 1817 Racquets Club by McKim, Mead and White on the other side of Park Avenue."[22] On the other hand, although some of Seagram's "crystallization" of and "differentiating" fit with its site may be in this way positive, this is not essential for fit to yet be a proper part of a work's appreciation. As *Tilted Arc* demonstrates, some of a work's fit with its site may be, or at least may be regarded by many appreciators to be, dislocating and even destructive. Some would contend that this is the kind of relationship that many modernist towers have to their sites.

To pursue the issue of a less positive or at least more controversial fit between work and site, however, it is more illuminating to consider postmodern architecture. Although the so-called postmodern movement seemingly embraces any departure from the modernist tradition, it is said by Robert A. M. Stern to involve three main features: historical allusion, ornamentation, and contextualism.[23] Of these three, exclusive attention to the first two is sometimes termed postmodernism of reaction and a focus on the last, postmodernism of resistance.[24] Frampton calls the latter "critical regionalism," for it stresses not only the immediate context, the site, but also local architectural forms, building techniques, and topography and even relevant cultural, social, and political factors.[25] By contrast, postmodernism of reaction seemingly emphasizes historical illusion and ornamentation almost at the expense of all else. A case illustrating both sides of postmodernism is that of the Johnson and John Burgee New England Life Building on Boston's Boylston Street. This structure, an example of postmodernism of reaction, has been described as "the party crasher on this block," an "arrogant intruder in the cityscape," and a "Jukeboxlike" tower with "arched windows, urns, columns, a formal forecourt and all manner of elements that show that its designers flipped through the history books."[26] The design caused such a public outcry that a second matching tower was canceled and its site given to Stern. In contrast to the Johnson and Burgee building, Stern's tower, it is said, "is designed for this site and for this site alone, its details recall the elements that make the older buildings of this neighborhood special, and its overall form takes into account the larger urban design needs of the entire neighborhood."[27] The Stern building, a case of postmodernism of resistance, is "molded so specifically to the demands of this particular site" and "weaves so comfortably into the Boston streetscape" that it is claimed to "save" the overall site.[28]

In light of the distinction between postmodernism of resistance and postmodernism of reaction, it seems clear that the former, in addition to being perhaps the more promising direction for postmodern architecture to take, presents no new difficulties concerning the importance of appreciating the fit between work and site in the appreciation of architecture. The careful attention to site given by the works such as the Boston Stern building makes them comparable in mode of appreciation to works such as Taliesin West and Falling Water. In contrast, however, postmodernism of reaction, as noted, seemingly emphasizes historical illusion and ornamentation to the exclusion of everything else and thus presents a serious challenge to the importance of fit in the appreciation of architecture. Less extreme examples of this side of the movement, such as the AT&T, can perhaps be treated on the model of *Tilted Arc*. For example, in creating the AT&T, Johnson is said to have been attempting to "counter" some of "the anonymity and muteness" of the site, which he attributed to the surrounding modernist towers, including his own Seagram tower. And thus the relationship of the work to the site can be seen as involving "irony, frivolity, or calculated shock value."[29] It certainly is an attempt, as is *Tilted Arc*, to "restructure, both conceptually and perceptually, the organization of the site," although perhaps more a humorous than a destructive attempt. One author refers to the AT&T as "a standup joke."[30]

Other more eclectic and more solemn works of postmodernism of reaction, such as certain classic creations by Robert Venturi and Michael Graves, pose more of a problem. Extreme examples, such as the Johnson and Burgee, have been criticized as "historical gamesmanship," "do-it-yourself history," and "Disneyland Classicism" that leaves one "floundering between kitsch...and a nostalgia for past grandeur."[31] In such cases what sense can be made of the aesthetic appreciation of the fit of the work with its site? It seems there are two alternatives. On the one hand, as suggested above with the AT&T, we can push the analogy with *Tilted Arc* as far as possible, experiencing exactly how such structures "restructure," "dislocate," "deconstruct," and even "destroy" their sites as a part of our appreciation of the work. After all, "party crashers" or "arrogant intruders" nonetheless have relationships, and often very interesting ones, with those on whom they crash or intrude. On the other hand, if a work of postmodernism of reaction really has *no* fit with its site, positive or negative, if its "fit" with where it stands really is comparable to *Autumn Foliage's* "fit" with its wall, then a more extreme alternative suggests itself. This is to deny such a work the status of the art form of architecture, to demote it, as it were, to simply a decorated building. This is suggested by some. For example, architect Romaldo Giurgola speaks against "impositions on the site" contending that the "making of clear connections with a cultural past and present is very different from the sophisticated playing" that "passes for architecture today" and is that in terms of which "a building becomes true architecture."[32] Others echo such sentiments, concluding that it "takes a creative act, not clever cannibalism, to turn a building into art."[33] However, I suggest that rather than endorsing this radical alternative, it is more fruitful to pursue the fit of works of architecture in another way.

Form follows function and fit follows function

Concerning the two previously discussed challenges posed by works of architecture, the issue of existence and the issue of location, together with the auxiliary question of fit, we considered as examples either modernist or postmodern office towers and certain works by Frank Lloyd Wright. Thus, it is fitting that a third challenge be introduced by reference to Louis Sullivan, for Sullivan is sometimes credited with the first significant instantiations of, if not the invention of, the high-rise office building, and Wright was his most famous apprentice. In "The Tall Office Building Artistically Considered," Sullivan wrote that "the shape, form, outward expression, design or whatever we may choose, of the tall office building should in the very nature of things follow the functions of the building." This is because:

> It is the pervading law of all things…,of all true manifestations of the head, of the heart, of the soul, that life is recognizable in its expression, that form ever follows function. This is the law…And thus, when native instinct and sensibility shall govern the exercise of our beloved art; when the known law, the respected law, shall be that form ever follows function; then it may be proclaimed that we are on the high-road to a natural and satisfying art, an architecture that will soon become a fine art in the true, best sense of the word, an art that will live because it will be of the people, for the people, and by the people.[34]

Sullivan's remarks dramatically proclaim that architecture is by its nature a functional art, and thus they remind us of the most obvious question posed by any work of architecture: the question of what its function is, what does it do? Moreover, the slogan "form follows function" sums up the significance of this question, which is that, as Sullivan put it, "the shape, form, outward expression, design or whatever we may choose," of a work of architecture should follow the functions of the work. Thus, although Sullivan himself was apparently of minor influence in the architectural developments that followed him, his words yet capture both the nature of architecture as an art form and the ideas that have dominated the art in the twentieth century. Not only are these ideas evident in works such as those of Wright and others directly influenced by Sullivan, but they illuminate the whole of the modernist movement, noted for its so-called "functionalist" buildings. For example, Sullivan's ideas are given concert expression by Mies van der Rohe:

> The office building is a house of work...of organization, of clarity, of economy. Bright, wide, workrooms, easy to oversee, undivided except as the undertaking is divided. The maximum effect with the minimum expenditure of means. The materials are concrete, iron, glass."[35]

And although Le Corbusier claimed that architecture goes beyond mere needs, he is more closely associated with the idea that the house is "a machine for living in."[36] Even postmodernism does not escape the themes to which Sullivan gave voice. In the design of the AT&T, Johnson returns to the tripartite division of base, shaft, and capital that Sullivan first justified on the grounds that outer form must reflect inner function.[37]

As with the other questions raised by works of architecture, the question of function is illuminated by comparison with different art forms. When we confront *Autumn Foliage* we do not ask what its function is, nor does it pose the question, for in an important sense such a work has no function. Of course, if we hold a theory of art such as Tolstoy's, then we might ask a question about a particular work's function as a way of asking about the function, if any, of art in general. Alternatively, if we are unclear about the exact nature of the kind of art we confront, then we might ask about the work's function as a way of asking what it does as a work of art – for example, whether or not it is representational and if so, what it represents. But aside from these kinds of questions, the question of the particular function of *Autumn Foliage* has no place in our appreciation of that work. By contrast, the question of the particular function of a work such as Falling Water or the AT&T is both posed by the work and relevant to our appreciation of it. It is essential in our appreciation of a work of architecture to come to the correct realizations about what it does – to know whether it is an office building or a temple, a fortress or a cathedral, a house or a mausoleum, all of which, it is worth noting, are functional categories. Moreover, here architecture differs not only from works such as *Autumn Foliage* but also from the earthworks and placement pieces with which it seemingly has much in common. Although works such as *Complex One* and *Tilted Arc* may raise the question of existence and certainly pose questions about their location and fit, they do not have a particular function any more than does *Autumn Foliage*.

There are, however, other art forms that are more functional in nature. Especially relevant to architecture is what might be thought of as its poor cousin: the public monument. Consider the statues of generals or the World War I (WWI) and II memorials that adorn many city parks; or the Vietnam Veterans Memorial in Washington, DC.[38] We may wish to withhold the term "work of art" from some of such pieces, but in many cases that would be a mistake and certainly so with the latter work. Of many of such works, especially some of the rather abstract WWI pieces, we appropriately ask questions concerning function as we might ask of *Autumn Foliage*, for example, questions about whether or not it is representational and if so, what it represents. However, unlike *Autumn Foliage* or even earthworks and placement pieces, a monument poses another question of function, one that is not about what it is or does as a work of art, but rather about what it does in addition to being and acting as art. In addressing this question we typically consider what war, battle, action, or individuals the piece commemorates, why it does so, and so on; in short, we are concerned with what cultural, social, or political functions the monument has. And the realization of these kinds of functions is a vital part of our appreciation of the piece. Without such realizations when appreciating, for example, the Vietnam Veterans Memorial we experience only a dark wall with a lot of names on it, and our appreciation is accordingly both impoverished and inappropriate. Here both comparison and contrast with architecture is useful. In comparison, these works bring out clearly the way in which our complete experience of works of architecture is dependent upon our appreciation of the functions they have independent of being and acting as art, what might be termed their nonartistic functions. This is not surprising in light of the above-noted openness in the borders between architecture and rest of the world.

More illuminating, however, is the contrast between monuments and memorial pieces and works of architecture. The key difference concerns the nature of the nonartistic functions and the means by which they are carried out. Monuments and memorials serve functions such as commemorating, honoring, venerating, and glorifying; their purposes are to inform, remind, induce, and inspire. Consequently, such pieces carry out their functions straightforwardly and on their surfaces, and typically in a direct representational or symbolic manner. The general with horse and sword or the black granite face of the Vietnam Veterans Memorial with its 58,132 inscribed names each directly denotes a particular war and those who served and sacrificed in it. There is nothing hidden, no other way in which or other place where the function is actually carried out, no alternative mechanism by which or inner place in which the real work is done. Thus, with such works there is no need to insist that form should follow function; typically the form itself is what carries out the function – function is embodied in form. However, works of architecture serve other kinds of nonartistic functions; they protect, shelter, and comfort, providing places in which to live, work, and worship. Given the nature of such functions, they must typically be carried out literally rather than simply symbolically. Even if a cathedral symbolizes the glory and the power of God, it must still be a house of, and thereby provide a place in which to, worship. This fact about the nature of the functions of architecture in turn explains the importance of the insistence that concerning architecture form should follow function. This is because when form does not directly embody function, as it does in the symbolic case, then it is seemingly possible for form to *not* follow function, but to, as it were, go its own way. And this possibility is accentuated by a second significant feature of architecture's functional nature, which is related to the way in which works of architecture, in contrast to monuments, typically carry out their functions: works of architecture have insides.

The fact that works of architecture, unlike almost any other works of art, have, by their natures, insides, inner space as well as outer place, seemingly follows directly from their functionality and the particular kinds of functions they perform, for such functions are typically performed *inside* the works. In fact, if a structure has no inside either because it has no function of the relevant kind, such as with most artistic earthworks or because its function is carried out exclusively on its outside, as with some Central American temple mounds, it is somewhat difficult to see the structure as a work of architecture at all. This "insidedness" of architecture means that, as an auxiliary question to that of function, works of architecture pose the question of the nature of their inner space. And unlike any of the other previously considered art forms, realizations about the insides of works of architecture are a significant dimension of their appreciation. We may speculate about what the inside of *Complex One* or even, with more difficulty, of *Autumn Foliage* is like or about what it looks like, but such speculation is not a significant part, if a part at all, of our proper appreciation of such works. However, realizations about, and if possible the direct experience of, the inside of, for example, Falling Water is an essential part of its proper appreciation. It is not surprising that books on architecture have diagrams and illustrations of the insides of structures as well as of their outsides. Nor is it surprising that there is something essentially frustrating about appreciating the exterior of, for example, a cathedral or a small country church and then discovering that it is locked, that we are barred from appreciating the inside – and thus unable to complete our appreciation of the work.

The appreciative significance of the "insidedness" of works of architecture has two related ramifications. First, it further elaborates the importance of "form follows function"; it makes explicit that this slogan typically spans the gap, as it were, between inside and outside, for, as noted, although function is typically carried out inside a work, much of its most appreciable form is outside. Second, it raises a further question of fit in the appreciation of architecture. In addition to the issue of the fit of the work with its site, there is also that of the fit of the inside of the work with its outside. The upshot of these two ramifications is that the question of fit is really a question of a three-way fit, among inside, outside, and site, and, since much of the fit is accomplished by form, this larger question of fit may be characterized as whether or not *fit follows function*.[39] Some examples illuminate aspects of this larger question. We previously considered cases of lack of fit between the outside of an architectural work and its site, such as with certain postmodern buildings. There are similar examples of lack of fit between outside and functional inside, and the most dramatic kinds of cases are the most revealing. Consider the disconcerting experience of entering an old city-center post office or a grand railroad station only to find that it has been gutted and filled with trendy new boutiques and restaurants. Such expediencies may be necessary to save the "building," but what is saved is only the shell, half a work of architecture that can no longer be fully appreciated due to the lack of any real fit between the outside and the functional inside. An equally disconcerting but in a sense reverse experience can be had in the National Gallery of Canada, where deep inside, surrounded by rooms of Canadian and international paintings, one can enter and appreciate the complete and completely intact interior of Ottawa's old Rideau Street Convent Chapel. This is the egg without the shell.

Function, location, existence: the path of appreciation

In the preceding three sections, we considered three challenges posed by works of architecture but not typically by other art forms: the question of existence, the question of location, and the question of function. Each requires us to come to realizations about and to thus make appreciable a certain dimension of works of architecture: the first, the very existence of the work, the second, the fit of the work with its site, and the third, the fit of the site and the outside of the work with its functional inside. The conclusions of these three sections can now be brought together to outline a general way of meeting these challenges. In the first section, the discussion of Tolstoy's concerns about art suggests that the core idea necessary for addressing the question of existence involves the functional nature of architecture. This suggestion is elaborated and reinforced in the following two sections. In the second section, the question of existence – to be or not to be – is expanded and deepened to that of to be or not to be here, and thus it is aligned with the question of fit. In the third section, the issue of fit is similarly expanded and deepened into a three-way fit, and the idea that form must follow function is developed so as to suggest that fit must follow function. The upshot is a proposition that Tolstoy, Sullivan, and others would have endorsed: that the functionality of works of architecture is the key to all three concerns. In short, realizations about the function of a work of architecture are among the most significant means by which to make appreciable, first, the fit of the inside of the work with its outside, second, the fit of the work to its site, and, last, the very existence of the work.

Demonstrating how the appeal to the function of a work of architecture facilitates the appreciation of its three-way fit and its existence is relatively easy with works such as Falling Water. Perhaps this is because Falling Water is the creation of Wright, who, as a disciple of Sullivan, apparently designed the work such that, in Sullivan's words, "the shape, form, outward expression, design or whatever we may choose" follows the function of the building. Thus, it is easily and straightforwardly appreciable in light of its function. After all, Falling Water is a weekend "cottage," a place for a leisurely retreat, for a return to nature. From this function its causal yet intimate blending of inner and outer spaces, its "almost total fusion" with the natural rocky and wooded site, and its very existence on that site in the first place all follow smoothly. Moreover, it is not surprising that many twentieth-century modernist works lend themselves to a similar treatment, as do classic earlier works, especially the religious and public buildings that make up much of the history of architecture. However, works that strongly confront us with their existence and their apparent lack of fit, such as some of the postmodern towers considered previously, are more difficult cases. As noted, Johnson's AT&T Building leaves us pondering the desirability of its existence and, as we saw at the end of the third section of this chapter, buildings of the postmodernism of reaction, such as Johnson's and Burgee's New England Life Building, leave many observers pondering their status as works of the art of architecture.

To establish the importance of function concerning such hard cases, consideration of another postmodern work is worthwhile. Especially illuminating is Michael Grave's fortress-like Portland Building.[40] Even more so than some other postmodern structures, this work has been questioned concerning both its existence and its status as architecture.[41] Initially the building strikes many observers as fantastical, if not preposterous. Along with other flamboyant features, it was designed to have two multistoried rectangular, false columns on the main facades, a representational sculpture, Portlandia, over the main entry, and numerous garlands adorning the side facades. In such a case, the appeal to function may not make the work more appreciable on all counts. However, it does somewhat enhance our appreciation to know that the structure is the Portland City Public Services Building. This realization helps us to appreciate, in light of the building's overall function, the role of the above-mentioned seemingly self-indulgent features. Thus, for example, we come to appreciate the columns as an invocation of public gates, the sculpture as a reinterpretation of a motif on the city seal, and the garlands as traditional symbols of welcome. In short, although the realization of function may not address all aspects of such cases, it yet appears to deepen and enrich appreciation of even the most difficult examples. And in the last analysis, approaching works in this way is perhaps more fruitful than attempting to exclude them from the art of architecture.

In light of these observations, it is now possible to draw some very general conclusions about the appropriate aesthetic appreciation of works of architecture: to outline in a general way what might be called the path of appreciation for a work of architecture; and in doing so, hopefully to shed some light on some of those things that are especially rewarding in the appreciation of architecture. Consider the following scenario. We confront a particular work; it proclaims "Here I stand," posing for us the three challenges of its existence, its location, and its function. It calls on us to consider, as it were, why is it standing at all, why is it standing here, and what is it, quite literally, doing here? A significant and proper part of our appropriate appreciation of it, therefore, must involve coming to realizations that will answer these questions, solve these problems. However, we cannot address these questions as we might address the questions posed by a work such as *Autumn Foliage*, that is, by responding to "Here I stand" with the naive question "Where do I stand?" and after determining the right spot proceeding to contemplate the work from that spot. Basically this is because, if what has been previously noted is correct, the key to all three questions posed by a work of architecture is the answer to the last, the question of its function. And the function of a work of architecture is not readily and straightforwardly, if at all, realizable by simply contemplating the work from a particular spot in front of it.

That the function of a work of architecture is not typically easily realizable simply by static contemplation relates to a number of the features of architecture noted in the preceding sections. For example, given that appreciation of architecture is not an insular experience and that a work's function is typically nonartistic, the function of the work is unlikely to be found simply in its contemplation. In this sense the function of architecture is unlike the function Tolstoy attributes to art in general. It is not a function that acts directly on the appreciator, and thus it is not realizable by simple static contemplation of the work by the appreciator. Rather it is the kind of function that the appreciator must realize by experiencing the function itself. Moreover, it is not completely adequate simply to know the function of the work and to contemplate it in light of that knowledge. Aesthetic appreciation involves realizations achieved in experiencing the work: to appropriately appreciate *Autumn Foliage* it is not sufficient to simply know that it is a representational work and what it represents; rather this knowledge must be essentially involved in our experience of it. Similarly, with architecture our realizations about the function of a work must be involved in our experience of it. Thus, it is not in fact sufficient just to know that a structure is, for example, a cathedral in order for its functionality to fully facilitate our appreciation of it as a work of architecture. Ideally, appropriate appreciation would involve the realization of its religious function by direct experience of that function in action, for instance, by experiencing, although not necessarily taking part in, a mass within the cathedral itself.

However, the main reason that the function of a work of architecture is not easily realizable by static contemplation is not simply that its function is non-artistic in nature nor that its function must be experienced in action. Rather it is that this nonartistic action normally takes place *within* the structure. The function of a work is typically performed and thus typically experienced inside the work. It follows that the appreciation of architecture is ideally a process, a path of appreciation that leads to the experience of the function inside the work. Such appreciation is in this way somewhat similar to the appreciation of a literary or dramatic thriller or to the appreciation of a symphony, each of which leads to a climax. Perhaps this is why architecture is sometimes called frozen music. However, with architecture the path of appreciation is not as fully specified by the work as it is in the case of these other art forms. Rather appreciators must make their own way from the initial confrontation with the work to the experience of the work's inner function. Typically, appreciators move from the outside to the inside: first, approaching the work from a distance, second, closing with it and perhaps circling it, and, only last, entering the work. And in one sense this is the natural path of appreciation in that on confrontation with a work the questions of existence, location, and function logically arise in that order. Thus, in approaching, we experience a work's existence, in closing and circling, we experience its outer form and its fit with its site, and, lastly, upon entering, we experience the fit between its outer and inner space and experientially realize its function. As with the thriller or the symphony, the climax comes near the end of the path of appreciation.

There is, however, a further complication concerning architecture's path of appreciation, somewhat of a paradox that contributes to the special nature and richness of the aesthetic experience of works of architecture. It is that although on confrontation with a work the questions of its existence, location, and function naturally arise in that particular order, in our experience of the work the key realizations for addressing these questions – that is, those realizations concerning its function – typically come last. As noted, the physical path of appreciation typically runs from outside to inside. Thus, while the physical path flows in one direction, the path of appreciative application of the realizations gained from following that physical path flows in the opposite direction. This latter path, in one sense the real path of appreciation, involves a series of realizations running from the work's function back to, one after the other, the fit of its inner with its outer space, the fit of the work with its site, and, lastly, the very existence of the work. In short, the aesthetic appreciation of a work of architecture is an experiential process that is not only not completed until the end is reached, but also not completed until that end is read back into the whole process such that the overall experience is thereby deepened, enriched, and completed. This kind of convoluted and somewhat self-reflective mode of aesthetic appreciation, although not unique to the experience of architecture, is a facet of what makes such experience especially rewarding.

Notes

1 I do not defend this notion of aesthetic appreciation here, but see my "Appreciating Art and Appreciating Nature," in Salim Kemal and Ivan Gaskell (eds) *Landscape, Natural Beauty and the Arts*, Cambridge, Cambridge University Press, 1993, pp. 199–227 (reproduced in this volume, Chapter 7) and "Between Nature and Art" (in this volume, Chapter 8).

2 Tom Thomson, *Autumn Foliage*, 1916, 10 ½ by 8 ½ inches, The National Gallery of Canada, Ottawa, Canada.

3 William Shakespeare, *Hamlet*, Act III, scene I, lines 56–88 [1601], in G. B. Harrison (ed.) *Shakespeare: Major Plays and the Sonnets*, New York, Harcourt, Brace & World, 1948, p. 626.

4 Ibid., p. 626.

5 Philip Johnson, the American Telephone and Telegraph Building, 1980–83, New York. I think that in general works of architecture are perhaps less well known than are those of many other major art forms. Consequently, throughout this chapter I attempt to employ examples that, due to either their location, historical significance, or notoriety, have attained widespread recognition. Also, in accompanying footnotes I indicate the locations and construction dates of all the works discussed in the chapter.

6 Frank Lloyd Wright, Falling Water (The Kaufmann House), 1935–37, Bear Run, Pennsylvania; Taliesin West, 1934–38, Phoenix, Arizona.
7 Leo N. Tolstoy, *What is Art?* [1896], Indianapolis, Bobbs-Merrill, 1960, see especially Chapter 1.
8 Ibid., p. 13.
9 Ibid., p. 15.
10 Ibid., p. 150.
11 Martin Luther, as quoted in Roland H. Bainton, *Here I Stand: A Life of Martin Luther*, New York, Mentor, 1955, p. 144.
12 Ludwig Mies van der Rohe, with Philip Johnson, Seagram Building, 1958, New York; Gordon Bunshaft (Skidmore, Owings and Merrill), Union Carbide Corporation Building and Chase Manhattan Bank, 1960, New York.
13 Le Corbusier [Charles Edouard Jeanneret], Villa Savoye, 1929–31, Poissy, France.
14 Richard Serra, *Tilted Arc*, 1983, Jacob Javits Federal Building Plaza, New York.
15 Richard Serra, transcript of the New York City hearing [1985], reprinted from *Harper's*, July, 1985, in Margaret Battin, John Fisher, Ronald Moore, and Anita Silvers (eds) *Puzzles About Art: An Aesthetics Casebook*, New York, St Martin's, 1989, p. 182.
16 Elizabeth C. Barker, "Artworks on the Land," *Art in America*, 1976, vol. 64, pp. 92–6 reprinted in Alan Sonfist (ed.) *Art in the Land: A Critical Anthology of Environmental Art*, New York, Dutton, 1983, p. 75. The three works are Robert Smithson, *Spiral Jetty*, 1970, Great Salt Lake, Utah; Michael Heizer, *Double Negative*, 1969–71, Virgin River Mesa, Nevada; *Complex One*, 1972–76, South-central Nevada. I discuss other ramifications of the relationships of such works to their sites in "Is Environmental Art an Aesthetic Affront to Nature?" *Canadian Journal of Philosophy*, 1986, vol. 16, pp. 635–50 (reproduced in this volume, Chapter 10).
17 Barker, op. cit., p. 79.
18 Serra, transcript of the New York City hearing, op. cit., p. 182.
19 Judge Dominick DiCarlo, transcript of the New York City hearing, op. cit., p. 183.
20 Ibid., p. 183.
21 Kenneth Frampton, *Modern Architecture: A Critical History*, New York, Oxford University Press, 1980, p. 189.
22 Ibid., p. 237.
23 Robert A. M. Stern, quoted in Ada Louise Huxtable, "The Troubled State of Modern Architecture," *New York Review of Books*, May 1, 1980, pp. 22–9.
24 See, for example, Steven C. Bourassa, *The Aesthetics of Landscape*, London, Belhaven, 1991, pp. 136–9, or Hal Foster, "Postmodernism: A Preface," in Hal Foster (ed.) *The Anti-Aesthetic: Essays on Postmodern Culture*, Port Townsend, Bay Press, 1983, p. xii
25 See Kenneth Frampton, "Towards a Critical Regionalism: Six Points for an Architecture of Resistance," in Foster, op. cit., pp. 16–30.
26 Paul Goldberger, "A Tale of Two Towers on Boston's Boylston Street," *New York Times*, January 24, 1988, p. (H)31. This excellent example and Goldberger's treatment of it were brought to my attention by Bourassa.
27 Ibid., p. (H)31.
28 Ibid., p. (H)31.
29 Mary McLeod, "Architecture," in Stanley Trachtenberg (ed.) *The Postmodern Moment*, London, Greenwood, 1985, p. 34.
30 Huxtable, op. cit., p. 26.
31 Goldberger, op. cit., p. (H)34, Huxtable, op. cit., p. 26 and Mcleod, op. cit., p. 42.
32 Romaldo Giurgola, "Architecture: More Than a Building," *Architecture Australia*, 1987, vol. 76, pp. 43–6; quoted in Bourassa, op. cit., p. 144.
33 Huxtable, op. cit., p. 26.

34 Louis Sullivan, "The Tall Office Building Artistically Considered" [1896], reprinted in Tim and Charlotte Benton (eds) *Form and Function: A Source Book for the History of Architecture and Design 1890–1939*, London, Open University Press, 1975, pp. 13–14. I apply Sullivan's ideas to other matters in "On Appreciating Agricultural Landscapes," *Journal of Aesthetics and Art Criticism*, 1985, vol. 43, pp. 301–12 (reproduced in this volume, Chapter 12).
35 Ludwig Mies van der Rohe, quoted in Frampton, *Modern Architecture*, op. cit., p. 163. Likewise Walter Gropius proclaims: "We want…an architecture whose function is clearly recognizable in the relations of its form." In general see the "First Proclamation of the Weimar Bauhaus," in H. Bayer (ed.) *Bauhaus; 1919–1928*, New York, Museum of Modern Art, 1938; reprinted in Larry L. Ligo, *The Concept of Function in Twentieth-Century Architectural Criticism*, Ann Arbor, University of Michigan Research Press, 1984, p. 12.
36 Le Corbusier, "The New Spirit in Architecture" [1924], reprinted in Benton, op. cit., pp. 132–3.
37 Sullivan, op. cit., pp. 11–13.
38 Maya Ying Lin and Jan Scruggs, Vietnam Veterans Memorial, 1982, Washington, DC.
39 I develop a related notion, that of functional fit, in "Reconsidering the Aesthetics of Architecture," *The Journal of Aesthetic Education*, 1986, vol. 20, pp. 21–7.
40 Michael Graves, City of Portland Public Services Building, 1980–83, Portland, Oregon.
41 For example, architectural critic Alan Colquhoun states: "What the building is saying, with a power and intensity that are almost unique and not at all banal, is that architecture, as it has come down to us from history, is now impossible." See Kurt Forster, Arthur Drexler, Vincent Scully, Alan Colquhoun, Allan Greenberg, Philip Johnson, and John Burgee, "The Portland Building," *Skyline*, January, 1983, p. 19.

14

LANDSCAPE AND LITERATURE

Hillerman's landscapes and aesthetic relevance

In this concluding chapter, I revisit some of the themes introduced and discussed throughout this volume: themes such as formalism, aesthetic relevance, scientific knowledge, functional descriptions, postmodern appreciation, as well as others. I do so in order to suggest an overall framework for the aesthetic appreciation of both our natural and human environments. I approach the construction of this framework by way of the landscape descriptions we find in literature, asking about the relevance of such descriptions to the aesthetic appreciation of the landscapes so described. I focus on one particular example: the landscape descriptions of contemporary mystery writer Tony Hillerman.

The Tony Hillerman Companion introduces Hillerman's work with the following remark: "For more than twenty years, Tony Hillerman has been painting with his words, a vivid landscape of the beautiful, but enigmatic American Southwest and its native peoples."[1] Hillerman's best-known works, of course, are modern-day, murder mystery novels featuring Navajo Tribal Policemen, Lieutenant Joe Leaphorn and Officer Jim Chee. However, it is no surprise that *The Hillerman Companion* mentions the landscape first and foremost. The desert landscape of the American Southwest is as much a player in Hillerman's novels as are Navajo Policemen Leaphorn and Chee. Hillerman himself observes: "Quite a few people read me for the landscape…They live in New York or some other place…and it reminds them of that great southern Utah and northern Arizona and Four Corners country."[2]

In this chapter I use what one Hillerman commentator calls "Hillerman's trademark desert setting" as an example for the study of landscape appreciation.[3] I consider Hillerman's vivid descriptions of the landscapes of the Four Corners country of Utah, Arizona, Colorado, and New Mexico, reflecting on how these descriptions enliven our aesthetic appreciation of the desert landscape itself. There can be little doubt that Hillerman's writings do in fact engender and enhance landscape appreciation. His landscape descriptions are universally praised by his reviewers for their power and appeal and characterized with words and phrases such as "overpowering," "breathtaking," "mesmerizing," "shimmering with realism," and "making the desert come alive."[4] One reviewer even extols them as "so evocative of the land around the Four Corners area...that even if you've never been there, you'll think you have!"[5]

The philosophical issue for which I use Hillerman's landscape descriptions as case studies is one of the central problems in aesthetics: the question of aesthetic relevance. As noted in Chapters 7 and 8, it is the question of what is relevant to the appropriate aesthetic appreciation of any particular object of such appreciation. Of course, aesthetic appreciation by its nature focuses on that which the object of appreciation presents to the senses. Thus, the question of aesthetic relevance is more precisely the question of what, if anything, of that which the object itself does *not* present to the senses is relevant to its appropriate aesthetic appreciation. One of the architects of modern aesthetic theory puts the issue as follows: "Is it ever 'relevant' to aesthetic experience to have thoughts or images or bits of knowledge which are not present within the object itself? If these are ever relevant, under what conditions are they so?"[6]

Hillerman's landscape descriptions bring the issue of aesthetic relevance sharply to a head. Consider this example:

> ...the sun was just dipping behind the Chuskas now. On the vast, rolling prairie that led away from the highway toward the black shape of Ship Rock every clump of sagebrush, every juniper, every snakeweed, every hummock of bunch grass cast its long blue shadow – an infinity of lines of darkness undulating across the glowing landscape. Beautiful. Chee's spirit lifted...[the] earth sloped away to the south – empty, rolling gray-tan grassland with the black line of the highway receding toward the horizon like the mark of a ruling pen. Miles to the south, the sun reflected from the windshield of a northbound vehicle, a blink of brightness. Ship Rock rose like an oversized, free-form Gothic cathedral just to the right, miles away but looking close. Ten miles ahead Table Mesa sailed through its sea of buffalo grass, reminding Chee of the ultimate aircraft carrier. Across the highway from it, slanting sunlight illuminated the ragged black form of Barber Peak, a volcanic throat to geologists, a meeting place for witches in local lore.[7]

Initially, we might have a mixed reaction to such a landscape description. On the one hand, much of the information contained in it seemingly cannot but enhance and enrich our aesthetic appreciation of the particular landscape it describes. Yet on the other hand, it is in large part comprised of "thoughts or images or bits of knowledge which are not present within" the landscape itself. Is any of this "outside information" really relevant to the appropriate aesthetic appreciation of the described landscape? Or is all of it perhaps relevant? How or how not? Why or why not?[8]

Classic formalism and postmodern landscape appreciation

To address such questions, it is useful to briefly reiterate some of the background to the issue of aesthetic relevance elaborated in Chapter 8. The above-noted classic formulation comes from the mid-twentieth century and thus from the heyday of modern analytic aesthetics. As such, it stands, somewhat as a fulcrum, between earlier formalist approaches to the problem and later postmodern approaches. As elaborated in Chapter 8, these two alternatives illustrate the extremes in answering the question of aesthetic relevance.[9] On the one hand, the classic formalist view is essentially purist, characteristically holding that only that which an object presents to an appreciator's senses is relevant to its aesthetic appreciation. All "outside information" is at best irrelevant and at worse harmful interference with pure aesthetic experience. On the other hand, there are the excesses of the postmodern approach, which seemingly holds that anything and everything that an appreciator brings to an object is relevant to its aesthetic appreciation, for in an important sense the object of appreciation is created anew in light of that which the appreciator contributes to it.

Of course, we live in an increasingly postmodern world. Thus, it seems that the relevance of Hillerman's landscape descriptions to our aesthetic appreciation of landscapes should pose no problem. Hillerman simply provides us with "evocative" and "mesmerizing" new "thoughts or images or bits of knowledge" about a landscape, and such thoughts, images, and knowledge, although "not present within" the landscape itself, yet enhance our appreciation of it. His descriptions, as noted, "shimmer with realism" and "make the desert come alive." In this way, they seemingly constitute resources in light of which each of us can recreate the landscapes of the Southwest for ourselves. Indeed, as also noted, these descriptions are "so evocative...that even if you've never been there, you'll think you have." The implication seems to be that Hillerman's landscape descriptions are so powerful and compelling that they allow us to create for ourselves and appreciate a landscape without even seeing it!

Should we then simply embrace the postmodern alternative and forget any concerns about aesthetic relevance? Should we simply sit back and enjoy Hillerman's assistance in creating anew the landscapes of the American Southwest? Such an approach is tempting, especially in light of the compelling and evocative nature of Hillerman's descriptions. However, it may be hasty. There are two considerations that should give us pause. The first is that, as noted in Chapter 8, we need not simply accept the either/or situation suggested by the two alternatives of formalism and postmodernism. There is, of course, a spectrum of possible positions concerning the question of aesthetic relevance running between these two extremes. Perhaps the correct view lies somewhere on this spectrum. Thus, I propose to consider the question of aesthetic relevance by beginning with formalism and, by means of examining a number of Hillerman's descriptions, moving toward postmodernism. How far such an examination takes us may indicate the most plausible position – and, as an added benefit, suggest somewhat of a general framework for landscape appreciation.

The second reason for not hastily embracing the postmodern alternative has to do with Hillerman's descriptions themselves. As noted, they are widely acclaimed for their power and appeal. Moreover, much of their power and appeal is a function of their richness and diversity. Hillerman's landscape descriptions are not simply one-dimensional; they operate on many different levels and incorporate many different themes. Consider again the passage quoted above. It mentions beauty, of course, but it also speaks of ragged black forms, undulating lines, long blue shadows, and slanting sunlight. It refers to sagebrush, juniper, snakeweed, and buffalo grass, to Ship Rock, Table Mesa, Barber Peak, and the Chuskas, and to highways, ruling pens, Gothic cathedrals, and aircraft carriers. And it alludes to the volcanic throats of geologists, the meeting places of Navajo witches, and the feelings and associations of Navajo Tribal Policeman Jim Chee. Given this richness and diversity, it is quite probable that although some of the themes of such a description are aesthetically relevant to the appreciation of the landscape described, others are not. Thus, I proceed from formalism toward postmodernism by examining the different kinds of descriptive themes that can be identified in a selection of passages from Hillerman's novels.

Formal descriptions and ordinary descriptions

I identify nine different themes in Hillerman's landscape descriptions, which I initially divide into three general groups: what I call, although I do not want anything in particular to depend upon these labels, the formal, the factual, and the cultural. First, consider the formal; by this I mean the component of these descriptions that focuses primarily on forms, that is, on the shapes, lines, and colors that are traditionally favored by formalists. For example, the previously quoted passage characterizes the landscape in terms of "ragged black shapes" and "undulating lines." Consider another illustration:

> North, over Sleeping Ute Mountain in Colorado, over Utah's Abajo Mountains, great thunderheads were reaching toward their evening climax. Their tops, reflecting in the direct sun, were snowy white and the long streamers of ice crystals blown from them seemed to glitter. But at lower levels the light that struck them had been filtered through the clouds over the Chuskas and turned into shades of rose, pink, and red. Lower still, the failing light mottled them from pale blue-gray to the deepest blue. Overhead, the streaks of high-level cirrus clouds were being ignited by the sunset. They drove through a fiery twilight.[10]

Concerning the formal component of this and similar descriptions, I suggest that two points are quite obvious. The first is in concert with classic formalism.

It is that, insofar as indicating such formal aspects of any landscape is possible, doing so seems clearly relevant to the aesthetic appreciation of the landscape in question.[11] For example, describing thunderheads from their top to their bottom in descending levels of color, from snowy white, through shades of rose, pink, and red, to mottled pale blue-gray and deep blue certainly draws our appreciation to a dimension of the landscape that is seemingly at the very centre of aesthetic experience. What could be more paradigmatic of the aesthetic appreciation of landscapes than the appreciation of a fiery sunset?

The second point, however, is contra formalism. It is that aesthetic appreciation of landscapes cannot be limited to this formal level. Classic formalism, as noted, is essentially purist, holding that only that which is presented to the senses is aesthetically relevant. Thus, the formalist typically limits aesthetic appreciation to the formal dimensions of objects of appreciation, contending that when and insofar as appreciation is informed by reference to anything beyond the formal it is no longer aesthetic in nature. However, as argued in Chapters 2 and 3, such purism cannot be maintained. This is particularly evident when we consider landscape descriptions that are as rich and as complex as are Hillerman's. In such descriptions it is clear that their formal components are inextricably intertwined with the second group of the three previously distinguished groups of descriptive themes, what I term the factual. To illustrate this, I turn to this component.

I suggest that at least four different descriptive themes can be distinguished within the factual component of Hillerman's landscape descriptions: the ordinary, the scientific, the historical, and the functional. By the first, the ordinary, I mean simply the characterization of landscapes with commonplace descriptive terms such as, in the previously example, "mountain," "cloud," and "sunset." This aspect of Hillerman's descriptions is sufficient to illustrate the point that aesthetic appreciation of landscapes cannot be limited to formal dimensions. Consider again the two previously quoted passages. In each Hillerman describes the basic features of a landscape as they are illuminated by a sunset. In considering these descriptions, it becomes clear that what is described and what we in our aesthetic experience of landscapes appreciate is not simply, for example, black shapes or shades of rose, pink, and red or fiery colors. Rather what is described and what we appreciate are the black shapes *of mountains*, rose, pink, and red *clouds*, and fiery *sunsets*. Hillerman's descriptions bring out the fact that the shapes, shades, and colors are inseparably connected to the mountains, clouds, and sunsets of which they are predicated. Indeed, this gives his descriptions a sense of solidity and concreteness, which contributes to the way in which they "make the desert come alive," and "shimmer with realism."

This line of argument is developed in more general terms in Chapter 2 and as such can be put as follows: we do not simply appreciate lines, shapes, and colors, but rather the lines, shapes, and colors *of* particular things. This is in part because the very identity and the specification of that which we appreciate at the formal level is in large measure a function of factual descriptions of at least the ordinary kind. We cannot access the former without the latter and, thus, the appreciation of the former brings with it the appreciation of the latter. Therefore, since the former are aesthetically relevant so too are the latter. Concerning the problem of aesthetic relevance, this means that at least mundane and commonplace "outside information" such as that we are viewing mountains, clouds, and sunsets is relevant to aesthetic appreciation, and is not, contra formalism, at best irrelevant and at worse harmful interference with pure aesthetic experience. In short, the dependence of the formal aspects of landscapes on ordinary descriptive terms takes us on the first step away from classic formalism and toward a position of the postmodern kind.

Other factual landscape descriptions

What about the other three themes that make up the factual component of Hillerman's landscape descriptions, that is, scientific, historical, and functional descriptions? Do they constitute irrelevant "outside information" that only interferes with aesthetic appreciation of landscapes, or are they, like ordinary descriptions, inextricably involved in such appreciation? I suggest that these other kinds of factual descriptions are, just as ordinary descriptions, importantly aesthetically relevant. This is in part because, since these different kinds of descriptions are all factual in a more or less straightforward sense, there is no hard and fast line that can be drawn between ordinary descriptions and the other three kinds. Ordinary descriptions blend into scientific, historical, and functional descriptions in such a way that the aesthetic relevance of the former necessarily bestows such relevance on the latter. If the former are aesthetically relevant so too are the latter.

In Hillerman's writing, the continuum from ordinary to the other kinds of factual descriptions is especially clear concerning scientific descriptions, that is, characterizations of landscapes in the terminology of the natural sciences.[12] For example, Hillerman's passages frequently involve geological characterizations of the landscape. Consider the following:

> The sun was almost on the horizon, shining through a narrow slot not closed in the west between cloud and earth. The slanting light outlined every crevasse in shadowy relief, making apparent the broken ruggedness of the ridge. It rose, ragged and tumbled shapes in black and gray, out of a long sloping hummock – what a million years or so of erosion had left of the mountain of volcanic ash that had once buried the volcano's core...It wasn't really a mountain. Technically it was probably a volcanic throat – another of those ragged upthrusts of black basalt that jutted out of the prairie here and there east of the Chuskas.[13]

This example illustrates the way in which Hillerman's descriptions blend the ordinary and the scientific. The geological characterization given in terms such as "erosion," "volcanic ash," "volcanic throat," and "basalt," is simply an extension of the ordinary description using terms such as "crevasse," "ridge," "hummock," and "mountain." The scientific description is more technical, somewhat finer-grained, and certainly theoretically richer than is the ordinary description, but it is not essentially different concerning its relevance to aesthetic experience. For purposes of aesthetic appreciation, a term such as "volcanic throat" performs the same function as a term such as "mountain," provided, of course, that the appreciator is familiar with geological terminology. Moreover, given such familiarity, insofar as the scientific description alters appreciation, it does not change it from aesthetic appreciation to something else, but rather only from a shallower to a deeper appreciation. In this sense scientific descriptions are not only not aesthetically irrelevant, they actually enrich and enhance appreciation. Indeed, Hillerman's frequent use of geological as well as biological, meteorological, and other kinds of scientific information in his landscape descriptions is one of the factors contributing to their widely acclaimed power and appeal.

Concerning the remaining two themes that constitute the factual component of Hillerman's landscape descriptions, that is, historical and functional descriptions, much the same can be said. They, like the scientific, blend with ordinary descriptions, and, rather than being irrelevant, actually enrich aesthetic appreciation. Consider the following example:

> The Indian Health Service Hospital at Gallup is one of the prides of this huge federal bureaucracy – modern, attractive, well located…the view from [Leaphorn's] window…was superb. The Health Service had located the hospital high on the slope overlooking Gallup from the south. Over the little hump in the sheet produced by his toes, Leaphorn could see the endless stream of semitrailers moving along Interstate 40. Beyond the highway, intercontinental train traffic rolled east and west on the Santa Fe main trunk. Above and beyond the railroads, beyond the clutter of east Gallup, the red cliffs of Mesa de los Lobos rose – their redness diminished a little by the blue haze of distance, and above them was the gray-green shape of the high country of the Navajo borderlands, where the Big Reservation faded into Checkerboard Reservation.[14]

Although this example is in many ways unlike the three previously quoted "sunset" passages, it is yet a landscape description of power and beauty. Its main difference from those descriptions is that it characterizes the landscape not with scientific terminology and not simply with formal and ordinary words and phrases. Rather it describes the landscape with historical and functional terms, using words and phrases such as "semitrailers," "Interstate 40," "highway," "intercontinental train traffic," "Santa Fe main trunk," "railroads," "east Gallop," "Navajo borderlands," and "Checkerboard Reservation." These terms capture the landscape in historical and functional categories, which are, of course, completely appropriate given the landscape in question. Moreover, given that landscape, this description, in moving beyond ordinary descriptions, performs the same service as do scientific descriptions, that is, it enriches and enhances appreciation. One need only consider contemplating the landscape of east Gallop and the hills beyond *without* the factual information contained in the functional and historical components of Hillerman's description to realize the extent to which this is the case.

The analogy with art argument

The contention that historical and functional landscape descriptions, as well as scientific ones, rather than being irrelevant, actually enhance aesthetic appreciation does not follow solely from the fact that descriptions in scientific, historical, and functional terms blend with those in ordinary terms. This contention also finds support in a line of argument that is implicit throughout much of this volume and may be called the Analogy with Art Argument. This line of argument involves two basic assumptions. The first is that at the most fundamental level the nature of aesthetic appreciation is essentially uniform across the appreciation of different kinds of objects and, thus, that at this level the aesthetic appreciation of landscapes and that of works of art mirror one another. The second basic assumption is that in the aesthetic appreciation of works of art certain kinds of historical, cultural, and artistic information is aesthetically relevant and enhances appreciation. The specific information concerns answers to questions such as why the work was produced, how it was produced, and why it was produced in the way in which it was. The relevance of this kind of information to the aesthetic appreciation of works of art is argued for in Chapter 8, where it is termed the "history of production" of the work. In short, then, the second assumption is that the history of production of any work of art is importantly aesthetically relevant.

It follows from the two premises of the Analogy with Art Argument that in the aesthetic appreciation of a landscape, the analogue of a work of art's history of production is aesthetically relevant and enhances appreciation. But what is a landscape's history of production? As noted in Chapter 8, at the basic level, this analogous history of production for any landscape is the information given in scientific descriptions of the formation and the development of the landscape, such as, for instance, Hillerman's geological descriptions. However, landscapes, somewhat unlike works of art, do not have a final completion point; they are inherently unfinished. Therefore, their ongoing development is also a part of their histories of production. This ongoing development is recorded not only in scientific descriptions but also in historical and functional descriptions, such as Hillerman's description of the landscape of east Gallop. The history of production of the landscape of Gallop continues into the present and is recorded in the descriptions of its recent history and its currents uses. Thus, historical and functional descriptions that tell of this history of production are aesthetically relevant and enhance aesthetic appreciation.

The way in which the history of production of a landscape is ongoing and in particular how historical and functional descriptions of it enhance aesthetic appreciation is especially evident in the case of agricultural landscapes. Consider the following example:

> On the north side of Highway 44, the ocean of sagebrush stretched away into the Angel Peak badlands. On the south side of the highway where the NAI [Navajo Agricultural Industries] held domain, the black-gray-silver of the sage had been replaced by mile after mile of green, the shade depending on the crop and the season. Dense stands of cornstalks alternated with thousands of acres of potato fields, followed by great circles of kelly green alfalfa, and incredible expanses of onions, watermelons, cantaloupes, cucumbers, sugar beets, whatever crop the market demanded. And all of this had been made possible by a rare and seemingly small Navajo victory over white land-grabbers…While the whites had taken nearly all the good bottom land, the Navajos still owned the water [of the San Juan River] and an infinity of worthless high desert hills. Now, from planting season until harvest, that water was showered out over the desert through elaborate mobile sprinkler systems. It turned the hills lush and green and produced jobs for hundreds of Navajos.[15]

This passage describes a particular agricultural landscape in a number of aesthetically relevant ways having to do with the function to which that landscape is currently put as well as the history that made this function possible. Much of the passage is a functional description in terms of the crops – corn, potatoes, alfalfa, onions, watermelons, cantaloupes, cucumbers, sugar beets – which the Navajo Agricultural Industries grow on the desert landscape by means of irrigation from the San Juan River. However, there is also reference to the history that made this project possible – a history of confrontation with and ultimately victory over white settlers who attempted to encroach on this land during the last century. Moreover, the way in which this history and use of the land has shaped this agricultural landscape is underscored by the contrast with the natural sagebrush landscape to the north of Highway 44. Such functional and historical descriptions record and recount the history of production of the landscape in question and thus, according to the Analogy with Art Argument, are not simply aesthetically relevant but also enhance and enrich the appreciation of that landscape.

Returning now to the general question of aesthetic relevance, in light of the foregoing observations, I tentatively conclude that mundane and commonplace information, such as that we are viewing mountains, clouds, and sunsets, is not all the "outside information" that is relevant to aesthetic appreciation. Equally relevant, and perhaps even more so if appreciation is to be rich and deep, is that we are contemplating, for example, a basalt volcanic throat exposed by a million years of erosion or intercontinental train traffic rolling east and west on the Santa Fe main trunk. Similarly relevant is information such as that we are viewing the Checkerboard Reservation in the high country of the Navajo borderlands or the expanses of onions, watermelons, cantaloupes, cucumbers, and sugar beets made possible by the irrigation projects of the Navajo Agricultural Industries. Such examples of "outside information," just as the more mundane and commonplace instances, are both factual and descriptive, that is, all such "outside information" involves factual information that describes the landscape in ways that influence how we see it. We might say that such information meets a descriptive condition, which we can formulate as follows: to be aesthetically relevant to the appreciation of a landscape, "outside information" contained in a landscape description must be descriptive of the actual landscape and descriptive in such a way that awareness of it affects and directs perception of that landscape. Such a descriptive condition might constitute part of a more general answer to the question of aesthetic relevance, for although it is clearly not sufficient for aesthetic relevance, it is seemingly a necessary condition. I return to this point below.

The conclusion that in light of their factual and descriptive natures all of the different components that we have so far considered are aesthetically relevant is a long way from classic formalism. Its assumption, as noted, is that all such information, with the exception of the formal, is at best irrelevant and at worst harmful interference with pure aesthetic experience. However, this conclusion is not yet close to the extravagances suggested by postmodernism. Moreover, perhaps we are here nearing the outer limits of relevant "outside information." However, to determine these limits and thus to see clearly the full extent of the move away from formalism, as well as to ultimately test the plausibility of the postmodern approach, I turn now to the remaining components of Hillerman's descriptions – those which I label cultural.

Nominal descriptions

Within the cultural component of Hillerman's landscape descriptions, as within the factual, at least four different descriptive themes can be distinguished: what I call the nominal, the imaginative, the mythological, and the literary. It should be noted that in general the cultural component of landscape descriptions frequently includes additional dimensions that are not clearly captured within these groupings, but these four themes are those most evident in Hillerman's descriptions and are, moreover, sufficient for present purposes.[16] The first, the nominal, refers to what are perhaps the most basic means by which we culturally characterize landscapes, the proper names with which we label them. Consider this example:

> Jimmie Yellow's place...seemed to have been selected more for the view than for convenience. It was perched near the rim of the mesa, looking down into the great empty breaks that fell away to the Rio Puerco. To the west, across the Laguna Reservation, the snowy ridges of Turquoise Mountain reflected the light of the rising moon. To the east, the humped ridge of the Sandia Mountains rose against the horizon, their base lit by the glowing lights of Albuquerque. To the north, another line of white marked the snowcap on the Sangre de Cristo Mountains, and the bright smudge of yellow light below them was Santa Fe, one hundred miles away. A spectacular view, but no water.[17]

Like the points of light that sparkle in the described view, this description itself is highlighted by proper names: not only "Jimmie Yellow," but also "Rio Puerco," "Turquoise Mountain," "Sandia Mountains," "Albuquerque," "Sangre de Christo Mountains," and "Santa Fe." How do such names function in aesthetic appreciation? In one sense names are typically very different from the previously discussed ordinary, scientific, historical, and functional descriptions. Names, unlike these descriptions, do not necessarily provide factual information about a landscape nor do they necessarily describe it. Thus, they do not necessarily enhance aesthetic appreciation. Rather they frequently act only as pointers, picking out and directing attention to certain landscape features. In this, they function like the most basic descriptions of landscapes, ones in which features are not identified by descriptive terms, such as "mountain," "cloud," and "sunset," but rather only with indexical words, such as "this" and "that." Such words, although they facilitate directing and focusing of aesthetic appreciation, do not, in that they have no descriptive content, contribute to it. The same is true of many proper names.

Not all proper names are without descriptive content, however. Many have such content in the sense that they have a specific meaning. For example, in the previously quoted passage, "Puerco" means pig, "Sandia," watermelon, "Sangre de Christo," blood of Christ, "Santa Fe," holy faith, and "Turquoise," of course, turquoise, although apparently "Albuquerque" has no such literal meaning. Likewise consider the names for the two butte's referred to in the following description:

> Now [Agnes Tsosie] was looking away, out across the gentle slope that fell away from Tesihim Butte and then rose gradually toward the sharp dark outline of Nipple Butte to the west. The sage was gray and silver with autumn, the late afternoon sun laced it with slanting shadows, and everywhere there was the yellow of blooming snakeweed and the purple of the asters. Beauty before her, Leaphorn thought. Beauty all around her.[18]

Compare the impact of the name "Nipple Butte" with that of the name "Tesihim Butte." The difference in impact makes evident that when proper names, such as "Rio Puerco," "Turquoise Mountain," "Sandia Mountains," "Sangre de Christo Mountains," "Santa Fe," and "Nipple Butte," and unlike "Albuquerque" or "Tesihim Butte," have meaning, then they have the potential to be descriptive of the landscape, just as do the descriptive terms of ordinary, scientific, historical, and functional descriptions. And when names are descriptive in this way, they, in meeting the previously indicated descriptive condition, can enhance aesthetic appreciation much as do factual descriptions. However, two qualifications to this claim need to be noted. First, the fact that a proper name has meaning does not guarantee that it is actually descriptive of the landscape that it names. The names for the Sandia Mountains and for Nipple Butte are in fact descriptive in the relevant sense, but it is not clear that the name for the Rio Puerco or for the Sangre de Christo Mountains or for Santa Fe is. However, if such names are in fact descriptive of what they name, seeing that they are and how they are can be a significant dimension of aesthetic appreciation.[19] Second, and more important, simply the fact that a name, or any other cultural attribution, is descriptive is not sufficient to enhance appreciation of a landscape. To be relevant to aesthetic appreciation, a cultural landscape description, in addition to meeting the descriptive condition, must also meet at least one other condition. To make this clear it is useful to consider the second of the cultural themes in Hillerman's descriptions, the imaginative.

Imaginative descriptions and cultural embeddedness

The imaginative theme of Hillerman's landscape descriptions is evident in the following example:

> Shiprock stuck up like a blue thumb on the western horizon seventy miles away. Behind it, the dim outline of the Carrizo Mountains formed the last margin of the planet. The sagebrush flats between were dappled with the shadow of clouds, drifting eastward under the noon sun.[20]

What I call the imaginative component is also well illustrated in the passage quoted at the outset of this chapter. Parts of that description read as follows: "Ship Rock rose like an oversized, free-form Gothic cathedral" and "Table Mesa sailed through its sea of buffalo grass, reminding Chee of the ultimate aircraft carrier."[21]

In these passages there are two proper names, Ship Rock and Table Mesa, both of which, due to their descriptive content, contribute, at least minimally, to the aesthetic appreciation of the landscape features they name. However, there are also three other descriptions of these same features, descriptions of the kind I label imaginative. These are the characterizations of Ship Rock as "a blue thumb" and "an oversized, free-form Gothic cathedral" and of Table Mesa as "the ultimate aircraft carrier." Such imaginative descriptions are equally as descriptive as, indeed even more so than, names such as Ship Rock and Table Mesa. However, the former do not, as the latter, contribute to appreciation, for they fail to meet a condition that is met by the latter. This is that in order to be aesthetically relevant, a cultural component of a landscape description must not simply be descriptive of the feature in question, but must also be what I term "culturally embedded" and thereby more generally available. Although it is difficult to state this embeddedness requirement precisely, it can be put somewhat as follows: to be aesthetically relevant to the appreciation of a landscape, cultural "outside information" contained in a landscape description must be embedded in a cultural framework and embedded in such a way that it is accessible within that particular framework and thereby accessible independent of the particular description. It is evident, I think, that proper names such as Ship Rock and Table Mesa are so embedded, while various possible imaginative characterizations, such as looking like a huge blue thumb or an oversized, free-form Gothic cathedral or as being reminiscent of the ultimate aircraft carrier, are not.

The justification for the embeddedness condition has to do with the way in which culturally embedded descriptions connect to the landscapes to which they apply. The previously considered factual descriptions, whether ordinary, scientific, historical, or functional, are aesthetically relevant and able to enhance appreciation because they are both descriptive and straightforwardly factual. Consequently, they connect to the landscapes they describe in such a way that, if they are true, then it is simply the case that the feature *is* that which it is described as. For example, Turquoise Mountain *is* a mountain, Ship Rock *is* a volcanic throat, and the rail line beyond Interstate 40 *is* the Santa Fe main trunk. This is, of course, part of what is meant by calling such descriptions factual. However, cultural descriptions, for instance, attributions such as proper names, can attain a certain kind of factuality if they are sufficiently culturally embedded. Thus, given the cultural embeddedness of the names "Ship Rock" and "Table Mesa," it is simply the case that Ship Rock *is* Ship Rock and Table Mesa *is* Table Mesa. These are, we might say, cultural facts. By contrast, Ship Rock is neither a blue thumb nor a Gothic cathedral and Table Mesa is not the ultimate aircraft carrier. To attribute such descriptions, we must use phrases such as "looks like," or "is reminiscent of." The importance of the embeddedness condition can be underscored by noting that, for instance, were "Thumb Rock" or "Cathedral Rock," rather than "Ship Rock," the culturally embedded name for Ship Rock, then it would simply be the case that Thumb Rock *is* Thumb Rock or that Cathedral Rock *is* Cathedral Rock and that it is only reminiscent of a ship. And the attributions that characterize the rock as a thumb or as a cathedral, and not that which likens it to a ship, would then be the ones that would be aesthetically relevant.

These observations concerning the nominal and the imaginative themes of

Hillerman's descriptions have obvious ramifications for the general question of aesthetic relevance. Of these two kinds of cultural attributions, the former enhance aesthetic appreciation if and when they meet each of two conditions, the descriptive condition and the embeddedness condition, while the latter, insofar as they fail to meet at least one of these conditions, are typically not aesthetically relevant. This suggests that these two conditions may provide a partial answer to the question of aesthetic relevance, the question, again, as the classic formulation puts it: "Is it ever 'relevant' to aesthetic experience to have thoughts or images or bits of knowledge which are not present within the object itself? If these are ever relevant, under what conditions are they so?" Perhaps such "outside information" is aesthetically relevant only if it, on the one hand, meets the descriptive condition and, on the other, provided that it is not straightforwardly factual in nature, meets the embeddedness condition as well as the descriptive condition. The two conditions are thus necessary, although, of course, not necessarily jointly sufficient for aesthetic relevance. If this is correct, it would mean we may be near the end of the slippery slope that begins with formal descriptions, continues through ordinary, scientific, historical, and functional descriptions, and leads now to the boundary between nominal and imaginary descriptions. It would also mean that the journey from classic formalism and toward postmodernism may likewise be near an end. These two conditions may constitute constraints on how much of a postmodern approach is acceptable. However, these conclusions are only tentative. To examine them more fully it is useful to turn to the remaining two cultural themes in Hillerman's descriptions, the mythological and the literary.

Mythological landscape descriptions

The mythological theme of Hillerman's landscape descriptions typically involves references to the rich Navajo mythological tradition. Consider the following example:

> It was cold here, frost still rimming the roadside weeds, and the snowcapped shape of the San Francisco Peaks twenty miles to the south looked close enough to touch in the clear, high-altitude air. The winter storm being held at bay by the Utah high in last night's weathercast was still hung up somewhere over the horizon. The only clouds this morning were high altitude cirrus so thin that the blue showed through them. Beautiful to Chee. He was back in Dine' Bike'yah, back between the Sacred Mountains, and he felt easy again – at home in a remembered landscape...This was one of those places for Chee – this desert sloping away to the hills that rose to become Dook'o'oosli'id, Evening Twilight Mountain, the Mountain of the West, the mountain built by First Man as the place where the holy Abalone Shell Boy would live, guarded by the Black Wind *yei.*[22]

Hillerman's use of Navajo mythology in his landscape descriptions is another factor that, like his use of scientific information, contributes significantly to their power and appeal. As illustrated in this passage, references to the four sacred mountains that indicate the four corners of Navajo Country, together with allusions to the mighty deeds performed there by the Navajo mythological heros, such as First Man, animate his descriptions. However, although such mythological dimensions obviously enrich the descriptions themselves, do they also enhance aesthetic appreciation of the landscapes described? Initially the answer to this question seems negative. On the one hand, mythological descriptions of landscapes are not factual, as are ordinary, scientific, historical, and functional descriptions, nor do they seem as substantially culturally embedded as are, for example, proper names. And on the other hand, with the exception of certain names that are parts of the mythology, such as "Tsodzil," that is, "Turquoise Mountain," they are seemingly not very descriptive of landscapes. Thus, at first glance it appears that the mythological component of these descriptions meets neither the embeddedness condition nor the descriptive condition.

Nonetheless, the conclusion that the mythological allusions in Hillerman's landscape descriptions are not relevant to the aesthetic appreciation of the landscapes in question seems somewhat paradoxical in light of their obviously significant role within the descriptions themselves. I suggest that this paradox may be resolved by considering what I call the scope of such mythological descriptions. These particular mythological descriptions are limited in scope in that, first, they, like proper names, apply to particular landscape features and, second, these features exist within a specific, circumscribed framework. The previously quoted mythological description is of one particular landscape – that of Dine' Bike'yah, the landscape between the Sacred Mountains. The specificity and the significance of this landscape is made evident in the following passage:

> Winding down the east slope of the Chuskas, Leaphorn stopped at an overlook. He pointed east and swept his hand northward, encompassing an immensity of rolling tan and gray grasslands. Zuni Mountains to the south, Jemez Mountains to the east, and far to the north the snowy San Juans in Colorado. "*Dinetah*," he said…"Among the People." The heartland of the Navajos. The place of their mythology, the Holy Land of the Dinee.[23]

What this quote brings out is that the mythological component of Hillerman's descriptions applies to a very particular landscape: the landscape that is the heartland of the Navajos. This is a landscape specified and limited both geographically and culturally. Moreover, it is, as the passage puts it, the Holy Land of the Navajos, the *place of their mythology*. Consequently, I suggest that concerning this specific framework, Hillerman's mythological descriptions are as deeply and as solidly culturally embedded as are proper names. In the same sense in which within a less limited framework Ship Rock simply is Ship Rock, within this geographically and culturally more limited framework, Evening Twilight Mountain simply is Evening Twilight Mountain. Moreover, since it is Evening Twilight Mountain, it is also the mountain built by First Man as the place where the holy Abalone Shell Boy would live. The point is that, provided they are attributed to particular landscape features, as are proper names, mythological attributions can be, within a specific framework, culturally embedded just as substantially as are proper names. Thus, in the same way in which the cultural fact that Ship Rock is Ship Rock enhances the aesthetic appreciation of the landscape feature so named, the cultural fact that Evening Twilight Mountain is the mountain built by First Man for Abalone Shell Boy may potentially enhance the appreciation of the specific landscape feature in question.

If a mythological description of a landscape can be, just as a proper name, adequately culturally embedded so as to open the possibility that it is aesthetically relevant, this still addresses only half the problem. There is yet the worry that, with the exception of names, such as "Turquoise Mountain," that may constitute a part of a mythological description, these so-called descriptions are seemingly not very descriptive of landscapes. To address this worry, it is useful to consider the ways in which some of the previously considered descriptions are in fact descriptive of the landscapes to which they are attributed. It may be argued that in one basic sense only formal descriptive terms, such as "turquoise," and some ordinary descriptive terms, such as "mountain," are straightforwardly descriptive of landscapes. However, even if this is the case, this basic sense clearly does not indicate the limits of aesthetic relevance. As noted previously, scientific, historical, and functional descriptions, all of which are descriptive of the landscape in a more extended sense, are aesthetically relevant. That a landscape feature is a basalt volcanic throat exposed by a million years of erosion or intercontinental train traffic rolling east and west on the Santa Fe main trunk is as aesthetically relevant as that it is a mountain or a train, or this or that color or shape.

If more complex scientific, historical, and functional factual descriptions adequately satisfy the descriptive condition, however, then there would seem to be no grounds for excluding mythological descriptions on the basis of this condition, provided, of course, that they meet the embeddedness condition. The comparison of mythological descriptions with scientific descriptions is especially revealing since both frequently involve, as an integral part of the description, an explanation of the origin or the nature of landscape features. For example, if the fact that a specific feature is the result of a million years of erosion is descriptive in the required sense, then so too is the cultural fact that it is the result of First Man's desire to accommodate Abalone Shell Boy. Indeed, Hillerman's landscape descriptions frequently illustrate this comparison. Consider these two parallel observations about the "Long Black Ridges" landscape surrounding Ship Rock. First: "...the volcanic action that formed Ship Rock lasted for tens of thousands of years. The pressure formed a lot of cracking in the earth's surface and every thousand years or so...there would be another bubbling up of melted rock and new ridges would form." And second: "According to Navajo mythology...those lava flows are the dried blood of the monsters killed by the Hero Twins."[24] Although we may judge the latter explanation to be less straightforwardly factual than the former, we are seemingly not justified in judging it any less descriptive of the landscape in question.

In light of these observations it seems plausible to conclude that mythological, as well as some other cultural descriptions of landscapes, at least within properly circumscribed frameworks, are aesthetically relevant and enhance aesthetic appreciation of landscapes in much the same manner as do scientific, historical, and functional descriptions. This conclusion greatly expands the quantity of aesthetically relevant "outside information." However, it does not yet get us to the postmodern view, which contends that anything and everything that an appreciator brings to a landscape is relevant to its appreciation, for the landscape is essentially created in light of that which the appreciator contributes to it. To test this contention it is useful to turn to the last of the cultural themes of Hillerman's landscape descriptions, the literary. This component is especially relevant to the postmodern view, for it may be

Illustration 10 Ship Rock and the Long Black Ridges, New Mexico.

argued that Hillerman's literature itself constitutes resources with which each of us can create the landscape for him or herself. It is even suggested, as noted at the outset of this chapter, that his novels make the landscape of the Southwest come alive to such a degree that an appreciator can seemingly create and appreciate these landscapes without ever actually experiencing them. Thus, the literary component of Hillerman's descriptions is a pivotal example for those who would promote a postmodern approach to landscape appreciation.

Literary landscape descriptions

As an illustration of the literary theme in Hillerman's landscape descriptions, consider the following passage, in which this component is interwoven with a number of previously discussed themes:

> Finding Joseph Joe proved simple enough...Chee rolled his patrol car southward across the San Juan bridge with the north wind chasing him, then west toward Arizona, and then south again across the dry slopes of snakeweed and buffalo grass toward the towering black spire of basalt that gave the town of Shiprock its name. It was the landmark of Chee's childhood – jutting on the eastern horizon from his mother's place south of Kayenta, and a great black thumb stuck into the northern sky during the endless lonely winters he spent at the Two Gray Hills Boarding School. It was there he'd learned that the Rock with Wings of his uncle's legends had, eons ago, boiled and bubbled as molten lava in the throat of an immense cinder cone. The volcano had died, millions of years had passed, abrasive weather – like today's bitter wind – had worn away cinders and ash and left only tough black filling. In today's bleak autumn light, it thrust into the sky like a surreal gothic cathedral, soaring a thousand feet above the blowing grass and providing – even at five miles' distance – a ludicrously oversized backdrop for Joseph Joe's plank and tarpaper house.[25]

As in other passages previously considered, this description focuses on Ship Rock and the surrounding landscape. On the one hand, this landscape feature is referred to in a number of different aesthetically relevant ways. It is described in formal and ordinary terms as "a towering black spire of basalt," it is characterized in scientific terminology as the remains of "molten lava in the throat of an immense cinder cone," and it is alluded to in one of its Navajo mythological guises as "the Rock with Wings." On the other hand, it is also characterized in imaginative ways similar to those discussed previously, as "a great black thumb" and as "a surreal gothic cathedral." As noted, these latter descriptions are not relevant to aesthetic appreciation. However, this description also has another dimension, the literary, for in it two literary characters, Navajo Tribal Policeman Jim Chee and one of his potential witnesses, Joseph Joe, enter the scene. We learn that Ship Rock "was the landmark of Chee's childhood – jutting on the eastern horizon from his mother's place south of Kayenta" and that it "stuck into the northern sky during the endless lonely winters he spent at the Two Gray Hills Boarding School." We also learn that it is "a ludicrously oversized backdrop for Joseph Joe's plank and tarpaper house." Are these descriptions of Ship Rock aesthetically relevant?

The question of the aesthetic relevance of the literary component in Hillerman's landscape descriptions is perplexing. On the one hand, that Ship Rock was the significant landmark of Jim Chee's childhood is seemingly no less factual than is the cultural fact that the lava flows surrounding it were formed by the dried blood of the monsters killed by the Hero Twins. Moreover, both characterizations are adequately descriptive. Thus, perhaps the former should be judged as aesthetically relevant as is the latter. On the other hand, such literary allusions seem similar to the imaginative characterizations of Ship Rock as looking like a great black thumb or a surreal Gothic cathedral. Thus, perhaps the literary component, as the imaginative, should not count as aesthetically relevant. In response to this quandary, I admit that in many ways literary descriptions are much like mythological ones. Indeed, we might contemplate a genus of cultural facts with two species, mythological facts and literary facts, and this proposal would seemingly find favor in certain approaches to the philosophy of art and literature. Nonetheless, I suggest that, in spite of their similarity to mythological descriptions, literary descriptions, just as imaginative ones, are not adequately culturally embedded to be deemed aesthetically relevant. To satisfy the embeddedness condition requires a cultural status that no literature, as long as it remains only literature, can easily attain. Consequently, although the literary "fact" that Ship Rock was a significant landmark in Navajo Tribal Policeman Jim Chee's childhood is a important fact *within* Hillerman's novels, it is not relevant to the aesthetic appreciation of this landscape feature as it exists in the real world.

This conclusion has far-reaching ramifications for the general significance of literature to the aesthetic appreciation of landscapes. If it is correct, it applies not simply to Hillerman's mystery novels, but to all those literary works, such as the novels of, to mention only two classic examples, Thomas Hardy and John Steinbeck, that embellish and enrich various landscapes with a host of fictional characters and events. As such, the conclusion strikes many who appreciate this literature as counter-intuitive, for it seemingly runs counter to their experiences of such literature. In this respect Hillerman's work constitutes an especially nice example. As noted at the outset of this chapter, Hillerman himself observes that many people read his novels for the landscape descriptions. Moreover, people not only read his novels for the descriptions, they also travel to the Southwest to experience for themselves the landscapes he describes. However, frequently these individuals do not want to experience simply Ship Rock and the lava flows surrounding it or only the fiery sunsets over the Chuskas or even only the Holy Land of the Navajos. They also want to experience the landscapes travelled by Lieutenant Joe Leaphorn and Officer Jim Chee; they want to see the places where Leaphorn pursued "thieves of time" and where Chee encountered "skinwalkers" or "the people of darkness."[26] Fodor's travel guide to the American Southwest even informs tourists that in the Four Corners area the "Navajo tribal lands are patrolled by the Navajo Tribal Police (the same force that employs the fictitious protagonists of Tony Hillerman's detective series: Officer Jim Chee and Lieutenant Joe Leaphorn)!"[27] In short, the idea that the literary component of landscape descriptions *is* aesthetically relevant to the landscapes described has both general significance and great intuitive appeal. Therefore, in light of this significance and appeal, the conclusion that it is *not* aesthetically relevant should not rest solely on the embeddedness condition. Thus, I return to the Analogy with Art Argument, from which this conclusion gets additional support.

The analogy with art argument again

Previously I employed the Analogy with Art Argument to demonstrate that a landscape's history of production, as recorded and recounted in scientific, historical, and functional descriptions of that landscape, is importantly aesthetically relevant, just as is the history of production of a work of art. Here I employ it to support the conclusion that literary descriptions of a landscape are *not* relevant to its aesthetic appreciation. In this regard, the relevant premise of the Analogy with Art Argument is the presupposition that at the most fundamental level the nature of aesthetic appreciation is essentially uniform across the appreciation of different kinds of objects of appreciation and, thus, that the aesthetic appreciation of landscapes and the aesthetic appreciation of works of art mirror one another. Thus, to pursue the ramifications of this line of thought, a literary description of a work of art is useful. Consider the following:

> Leaphorn spent the rest of the afternoon prowling through the Museum of Modern Art. He sat, finally, where he could see the patio of sculpture, the rain-stained wall behind it, and the rainy sky above,...and watched the water run down the bricks, drip from the leaves, form its cold pools on the flagstones, and give a slick shine to Picasso's goat...The goat was Leaphorn's favorite. When they were young and he was attending the FBI Academy, he had brought Emma to see New York. They had discovered Picasso's goat together...Emma had laughed... and said: "Look. The mascot of the Navajo Nation."... "Perfect for us Dineh"..."It's starved, gaunt, bony, ugly. But look! It's tough. It endures."...And of course, it was true. That gaunt goat would have been the perfect symbol. Something to put on a pedestal and display. Miserable and starved, true enough. But it was also pregnant and defiant – exactly right to challenge the world at the

> entrance of the ugly octagonal Tribal Council meeting hall at Window Rock. Leaphorn remembered their having coffee at the museum cafe and then walking out and patting the goat. The sensation came back to him now – wet, cold metal slick under his palm – utterly real.[28]

In this passage Hillerman's most famous literary character, Navajo Tribal Policeman Lieutenant Joe Leaphorn, contemplates an equally famous real work of art, Picasso's *She-Goat* (1950).[29] He recalls the initial reaction that he and his late wife, Emma, had to the work. The incident described is both touching and significant within Hillerman's novel. Moreover, the description itself contains a number of interesting observations about the work of art. Some of these observations are aesthetically relevant. For example, it is clearly aesthetically relevant that the work, when in the rain, is shiny, cold, and slick to the touch and that the goat is represented, if it is, as starved, gaunt, bony, ugly, and yet pregnant and defiant. It is likewise relevant that it expresses toughness and endurance. Moreover, we might also accept the relevance of the observation that the work could be a perfect symbol for the Dineh, a mascot for the Navajo Nation. And perhaps we might even consider relevant the suggestion that it would pose exactly the right challenge to the world were it to be put on a pedestal and displayed at the entrance of the Tribal Council meeting hall at Window Rock. However, these last two observations reach and probably exceed the limits of aesthetic relevance. Seemingly they are more like the imaginative component of Hillerman's landscape descriptions and not in fact aesthetically relevant. Be that as it may, however, what is clearly beyond the limits of aesthetic relevance is the literary fact that Picasso's work of art is a favorite of Navajo Tribal Policeman Joe Leaphorn and that he and his wife, Emma, when they were young and he was attending the FBI Academy, discovered it together.

As this example illustrates, with works of art the outer limits of aesthetic relevance are, although somewhat fuzzy, yet clear enough to undoubtedly exclude the literary component of a passage such as Hillerman's description of Leaphorn's encounter with Picasso's goat. This conclusion is reinforced by the following thought experiment: imagine a volume of art history and art criticism on Picasso's sculpture.[30] Were such a study to describe Picasso's goat only in term of its formal and representational features, we would find it quite properly scholarly, although perhaps a bit boring. Were it to characterize the work in terms of its expressive and symbolic features, we might find it somewhat more interesting; and were it to then go on to consider various potential symbolic uses the piece could have, we might find it, although imaginatively interesting, perhaps somewhat too imaginative. However, were it to go even further and point out that *She-Goat* is one of Joe Leaphorn's favorite pieces, we would find it absolutely crazy! The ramifications for the appreciation of not simply art but also of landscapes are clear enough: since the literary fact that Picasso's goat is one of Navajo Tribal Policeman Joe Leaphorn's favorite works of art is clearly not aesthetically relevant to that work's appreciation, it follows by the Analogy with Art Argument that the literary fact that Ship Rock was an important landmark of Navajo Tribal Policeman Jim Chee's childhood is likewise aesthetically irrelevant to the appreciation of that landscape feature.

Conclusion

The literary theme is the last of the nine descriptive themes I distinguish in Hillerman's landscape descriptions. Thus, its consideration marks the end of our examination of aesthetic relevance and of the path running from formalism to postmodernism. Concerning the nine themes, I summarize as follows: Formal descriptions together with ordinary descriptions are clearly aesthetically relevant. Other factual descriptions, such as the scientific, the historical, and the functional, are also relevant to and enhance and enrich aesthetic appreciation. Similarly relevant, at least within their properly circumscribed frameworks and provided that they are adequately culturally embedded, are cultural descriptions such as the nominal and the mythological. What is excluded as aesthetically irrelevant "outside information" are cultural descriptions such as the imaginative and the literary. These conclusions not only demonstrate that classic formalism provides much too restrictive an answer to the question of aesthetic relevance, but also suggest that not just anything and everything that an appreciator happens to bring to a landscape is relevant to its aesthetic appreciation. Thus, these conclusions also limit the extremes to which postmodern landscape appreciation can reasonably be taken. The limits, as suggested earlier, are roughly marked by the descriptive and the embeddedness conditions. What these conditions exclude from the sum total of aesthetically relevant information are exactly those imaginative and literary allusions frequently favored by the postmodern point of view.

I conclude with one more landscape description from Hillerman – one that brings together much of that which is aesthetically relevant not simply for appropriate, but also for deep and rich appreciation of the landscape described:

> The sun was in its autumn mode, low in the southwest, and shadows slanted away from every juniper. They formed zebra stripes where the slopes ran north and a polka-dot pattern where they slanted. The grass was never really green in this land of little rain. Now it was a golden autumn tan with streaks of silver and white where the sickle-shaped seeds of grama were waving, tinted blue here and there by distance and shadow. Miles away, beyond the hills, the vertical slopes of Chivato Mesa formed a wall. Above the mesa stood the serene blue shape of Tsodzil, the Turquoise Mountain which First Man had built as one of the four sacred corner posts of Navajo Country. And over all that, the great arching multilayered sky – the thin translucent fan of ice crystals still glittering in the full sun. Thousands of feet

lower, a scattering of puffy gray-white cumulus clouds – outriders of the storm the weatherman had been predicting – marched eastward ahead of the wind.[31]

Notes

1 Martin Greenberg (ed.) *The Tony Hillerman Companion: A Comprehensive Guide to His Life and Work*, New York, Harper, 1994, endpiece.
2 Tony Hillerman, quoted in Brandon Griggs, "Hillerman Trades Southwest for Vietnam," *The Salt Lake Tribune*, November 5, 1995, p. E3.
3 Griggs, ibid., p. E3.
4 These phrases are gathered from a number of different reviews of Hillerman's novels.
5 An anonymous reviewer for the *Chicago Tribune*, quoted in Greenberg, op. cit., p. i.
6 Jerome Stolnitz, *Aesthetics and Philosophy of Art Criticism: A Critical Introduction*, Boston, Houghton Mifflin, 1960, p. 53. In his formulation of the question, Stolnitz uses the traditional idea of "aesthetic experience." In contrast, I employ the notion of "appropriate aesthetic appreciation," although to avoid this rather cumbersome phrase, I typically write only "aesthetic appreciation" or even only "appreciation." I prefer the notion of "appropriate aesthetic appreciation" to that of "aesthetic experience," for the former makes more explicit the implications both that aesthetic experience is active rather than passive and that it can, concerning an object of appreciation, be appropriate or not. I discuss these issues as well as Stolnitz's position and the question of aesthetic relevance more fully in "Appreciating Art and Appreciating Nature," in S. Kemal and I. Gaskell (eds) *Landscape, Natural Beauty and the Arts*, Cambridge, Cambridge University Press, 1993, pp. 199–227 (reproduced in this volume, Chapter 7) and in "Between Nature and Art" (in this volume, Chapter 8).
7 Tony Hillerman, *Coyote Waits*, New York, Harper, 1990, pp. 101–2.
8 In addition to these questions concerning the aesthetic appreciation of the landscape described in such a passage, there are also, of course, questions about the aesthetic appreciation of the description itself. I take these to be distinct issues in the aesthetics of literature and do not address them here.
9 I spell out these matters in more detail in "Between Nature and Art" (in this volume, Chapter 8).
10 Hillerman, *Coyote Waits*, op. cit., pp. 103–4.
11 I discuss the problem of indicating the formal qualities of landscapes as well as other problems with formalism concerning the natural environment in "Understanding and Aesthetic Experience" (in this volume, Chapter 2) and in "Formal Qualities in the Natural Environment," *Journal of Aesthetic Education*, 1979, vol. 13, pp. 99–114 (reproduced in this volume, Chapter 3).
12 I consider in greater detail and give additional arguments for the aesthetic relevance of scientific descriptions to landscapes in the first part of this volume, especially in "Appreciation and the Natural Environment," *Journal of Aesthetics and Art Criticism*, 1979, vol. 37, pp. 267–76 (reproduced in this volume, Chapter 4) and in "Nature and Positive Aesthetics," *Environmental Ethics*, 1984, vol. 6, pp. 5–34 (reproduced in this volume, Chapter 6). I also address these issues in "Saito on the Correct Aesthetic Appreciation of Nature," *Journal of Aesthetic Education*, 1986, vol. 20, pp. 85–93 and in "Nature, Aesthetic Appreciation, and Knowledge," *Journal of Aesthetics and Art Criticism*, 1995, vol. 53, pp. 393–400.
13 Hillerman, *Coyote Waits*, op. cit., pp. 315–17.
14 Tony Hillerman, *Skinwalkers*, New York, Harper, 1986, pp. 173–4.

15 Tony Hillerman, *Sacred Clowns*, New York, Harper, 1993, pp. 224–6. I discuss the role of functional descriptions in the aesthetic appreciation of agricultural landscapes in more detail in "On Appreciating Agricultural Landscapes," *Journal of Aesthetics and Art Criticism*, 1985, vol. 43, pp. 301–12 (reproduced in this volume, Chapter 12).

16 Other kinds of cultural characterizations of landscapes, such as, for example, the more general symbolic significance given to certain landscapes or the much-discussed "sense of place," are, of course, interesting cases for consideration. However, I think that much of what can be said of such cases is similar to what can be said concerning nominal and mythological descriptions. I think that the role of nominal descriptions in aesthetic appreciation is especially revealing, and yet it is almost totally ignored. For excellent backgrounding on this topic, see George R. Stewart's classic study, *Names on the Land: A Historical Account of Placenaming in the United States* [1945], San Francisco, Lexikos, 1982.

17 Tony Hillerman, *The Ghostway*, New York, Harper, 1984, pp. 262–3.

18 Tony Hillerman, *Talking God*, New York, Harper, 1989, p. 29.

19 The way in which a descriptive name enhances aesthetic appreciation of a landscape differs depending on the descriptive content of the name in question. On the one hand, seeing that Turquoise Mountain, for example, is appropriately named and how it exemplifies its name is seemingly a dimension of its aesthetic appreciation somewhat analogous to appreciation in light of formal descriptions. On the other hand, if knowledge of a name such as "Santa Fe" enhances appreciation, the way in which it does is perhaps more similar to the enhancement affected by historical and functional descriptions of the landscape. Santa Fe was named by Don Juan de Onate who arrived in the area in 1598 with the power from King Philip of Spain "to conquer and hold." Stewart, op. cit., p. 26, reports:

> As his capital Onate founded a city in a high valley between mountains. To fit with its dignity and his own love of pomp, he named it sonorously La Villa Real de la Santa Fe de San Francisco, 'The Royal City of the Holy Faith of Saint Francis,' but it was shortened to Santa Fe.

20 Tony Hillerman, *A Thief of Time*, New York, Harper, 1988, p. 148.

21 Hillerman, *Coyote Waits*, op. cit., p. 102.

22 Hillerman, *The Ghostway*, op. cit., pp. 224–5.

23 Hillerman, *Coyote Waits*, op. cit., pp. 202–3.

24 Ibid., pp. 105–6.

25 Hillerman, *The Ghostway*, op. cit., pp. 44–5.

26 These quotes are derived from the titles of the following: Hillerman, *A Thief of Time*, op. cit., Hillerman, *Skinwalkers*, op. cit., and Tony Hillerman, *People of Darkness*, New York, Harper, 1980.

27 Nancy Zimmerman and Kit Duane (eds) *The American Southwest*, Oakland, Fodor's Travel Publications, 1996, p. 158.

28 Hillerman, *A Thief of Time*, op. cit., pp. 192–3.

29 Pablo Picasso, *She-Goat*, 1950, Bronze, after found objects, 46 3/8 by 56 3/8 by 27 3/4 inches, Museum of Modern Art, New York.

30 I have in mind works such as, for example, Roland Penrose's classic *The Sculpture of Picasso*, New York, Museum of Modern Art, 1967 and Werner Spies's *Picasso Sculpture*, London, Thames and Hudson, 1972. Interestingly enough, critics do not seem very excited about the features of *She-Goat* that impress Joe and Emma Leaphorn. Critical comments tend to focus either on the fact that the work is "almost entirely composed of found objects" (Spies, ibid., p. 179) or that the goat has "grotesquely detailed sexual organs." See, for example, Alan Bowness, "Picasso's Sculpture," in Roland Penrose and John Golding (eds) *Picasso in Retrospect*, New York, Praeger,

1973 (the last quote in the previous sentence is from p. 153). Moreover, such observations as there are that are of roughly the same nature as those of Joe and Emma seem to contradict theirs, describing the goat as "docile" rather than "defiant." See, for example, Roland Penrose, *Universe Sculpture Series: Picasso*, New York, Universe Books, 1961 (the short quotes in the previous sentence are from p. 12).

31 Hillerman, *Sacred Clowns*, op. cit., p. 131. Earlier versions of this chapter were presented at The International Association for Philosophy and Literature, Irvine, California, May, 1998; The Association for the Study of Literature and Environment, Missoula, Montana, July, 1997, The Canadian Society for Aesthetics, St. John's, Newfoundland, June, 1997, The American Society for Aesthetics, Rocky Mountain Division, Santa Fe, New Mexico, July, 1996 and at several universities in New Zealand and Australia during April and May of 1997. I thank those present, as well as Arlene Kwasniak and Alex Neill, for valuable suggestions.

INDEX

Note: page numbers in **bold** indicate chapters